Answers to activities and assessments are available online at:

www.leckieandleckie.co.uk/page/Resources

Introduction

About this book

This book provides a resource to support you in your study of physics. The book covers the Key areas in the Units of National 3 and National 4 Physics. It also covers all the physics Experiences and Outcomes (Es and Os) for Curriculum for Excellence (CfE) Science at Third and Fourth Levels.

Companion biology and chemistry books provide the same support if you are studying biology or chemistry in any of these courses.

The book has been organised to match to the course specifications and provides examples, activities, exercises and features to deepen your understanding of physics and help you prepare for N3 and N4 course assessments.

The topics have been organised into three Units: Electricity and energy; Waves and radiation, and Dynamics and space. Each chapter has been broken down into chapters covering specific subject areas aligned with the course specifications.

Features

This chapter includes coverage of
Each chapter begins with a brief listing of the topics covered in the chapter, using the N3 and N4 and CfE subject descriptions and codes.

> **GO! This chapter includes coverage of:**
>
> N3 Electromagnetic radiation • N4 Electromagnetic spectrum • Forces, electricity and waves SCN 3-11b • Forces, electricity and waves SCN 4-11b

You should already know
After the list of topics covered in the chapter, there is sometimes a list of the topics **you should already know** from previous study. Some of these topics will have been covered in earlier years at school, while others may depend on other chapters in this book.

> **You should already know:**
> - that any object which orbits a planet or a star is called a satellite
> - that the Moon is a natural satellite orbiting the Earth
> - that the Earth is a natural satellite orbiting the Sun.

Scotland's leading educational publishers

S1 to National 4

PHYSICS

STUDENT BOOK

Anna Lee • James Spence

CONTENTS

Learning intentions

Each main section within a chapter has a set of **Learning intentions**. These are organised according to the N3 and N4 requirements and the CfE Levels, so you can see if a section is suitable for your part of the course you are studying. For example, if you are studying for N3 Physics, you don't need to read the N4 sections, and if you are studying for N4 Physics, you should already know what is in the N3 sections.

To help you navigate your way through the book, different colours are used to indicate content from the different CfE Levels and N3 and N4 qualifications. This colour-coding is also used in the footer bars at the bottom of each page and in the Learning checklist at the end of each chapter.

Uses of sound

Learning intentions	National 3
In this section you will:	National 4
	Curriculum level 4

Learning intentions

In this section you will:

- learn that there are different applications of sound waves
- investigate sound reproduction technologies.

	National 3
	National 4
	Curriculum level 4
	Forces, electricity and waves: Vibrations and waves SCN 4-11a

Activities

Each chapter includes a range of **activities** used to help develop your understanding and to assess your knowledge. There are activities to work on individually (☻) and other which encourage collaboration with a partner (☻☻) or in a group (☻☻). There are three different types of activities:

The ![teacher] graphic shows that an activity may need input from a teacher.

The ![warning] graphic highlights a general safety alert or indicates a stage in an experiment when particular care needs to be taken.

- experiments, which can be used to familiarise you with the apparatus and understand ideas and experimental methods. The experiments includes a list of things needed for the experiment, instructions explaining how to carry out the experiment, as well as questions that the experiments will raise. Most experiments will be carried out as group activities.

GO! Activity 9.2

1. Experiment In this activity you will use a sheet of paper and a swinging source of sand or paint to produce and label waves.

Read the plan carefully then follow the instructions, making sure you follow all of the safety rules as you would normally be expected to do when carrying out practical work.

You should work in a small group but record your own observations and complete the answers yourself.

You will need: a paper cup, a clamp and stand, a pin, string, some tape, fine sand and a long sheet of paper. (You can use diluted paint instead of sand.)

- research and discussion activities. These involve you carrying out a piece of research on a particular topic and preparing a report or discussing your findings with a partner or the rest your group or class.

GO! Activity 13.2

☻ Search online to find out the most radioactive places on Earth. Write a short report. Try to find out levels of background radiation where you live and compare them with the most radioactive places.

- Factual recall and numerical calculations. The calculations use equations and formulae to test your knowledge and understanding of some of the key calculations needed in the study of physics.

GO! Activity 9.5

☻ 1. Draw a wave shape and label a crest, a trough, the amplitude and a wavelength.

2. The wave in the diagram has a total height of 2.5 m. Calculate the amplitude.

3. The wave in the diagram has a total length of 12 m. Calculate the wavelength.

4. If 20 waves occupy a total space of 100 m, calculate the wavelength of one wave.

5. A guitar string is plucked and starts to vibrate. The total length of the string is 0.6 m and there are 2 complete waves along the string. Calculate the wavelength of the waves.

Examples

New topics involving calculations are introduced with at least one **example**. The examples show how to go about tackling the questions and activities. Each example breaks the question and solution down into steps, so you can see what calculations are involved, what kind of rearrangements are needed and how to work out the best way of answering the question.

Example 9.2

250 waves are produced in 20 seconds. Calculate the frequency.

number of waves = 250

time taken = 20 s ●━━━━━━━━(Write out what you know from the question.)

frequency = ?

$$\text{frequency} = \frac{\text{number of waves}}{\text{time taken}}$$ ●━━━(Write the equation (relationship).)

Highlight boxes

Highlight boxes are used to emphasise important equations.

speed = frequency × wavelength

This can be written as

$$v = f\lambda$$

where:

v is used to represent speed

f is used to represent frequency

λ is the Greek letter lambda and is used to represent wavelength.

Physics in action

Many things in the world rely on physical laws. **Physics in action** boxes show real-life examples of physics and the way it plays a part in the way we live our lives, and how it helps us explore our place in the Universe.

✴ Physics in action: Scientific endeavours in low Earth orbits

The Hubble Space Telescope was launched into a low Earth orbit in 1990. It can take images in the ultraviolet, visible and infrared regions of the EM spectrum. The telescope is over 13 metres long, and the mirror used to capture the light is 2.4 m in diameter.

The Hubble telescope has been responsible for some significant discoveries. It has helped to improve estimates for the age of the universe, and has taken some iconic images. The Hubble Ultra Deep Field image, in Figure 16.6, has revealed galaxies billions of light years away.

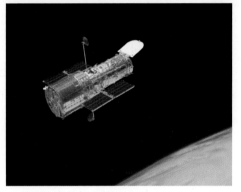

Figure 16.5 *The Hubble Space Telescope orbiting the Earth.*

Word banks

Word banks give short, simple explanations and definitions of important words and phrases.

📖 Word bank

• **Orbit**

An orbit is the curved path of an object as it revolves around a planet or star. Objects stay in orbit due to gravity.

Did you know ...?

Did you know boxes give you little snippets of information about physics and physicists, and where they appear in our lives and in the world around us.

❓ Did you know ...?

The image formed on the retina in our eye is upside down. Our brain has to decode the signal sent by the optic nerve, flipping it upright.

Hint

Hints are usually designed to help you with a particular activity or task.

Hint

Look back at Table 13.1. Which radiation is not stopped by air, paper or aluminium?

Make the link

Lots of areas of physics are related to other areas of physics and to other subjects as well. **Make the link** boxes shows how different aspects of physics link to each other and to other subjects.

Make the link

Compare the reflection of light off a smooth mirror with the reflection of sound from smooth and rough surfaces in Chapter 10.

Keep up to date!

Physics is a subject which advances all the time. **Keep up to date** boxes show topical points for discussion which you can explore with your own research.

♻ Keep up to date!

Partial solar eclipses take place fairly frequently. Search online to find out when the next partial solar eclipse will be visible in Scotland.

Learning checklist

Each chapter closes with a summary of learning statements showing what you should be able to do when you complete the chapter. You can use the Learning checklist to check you have a good understanding of the topics covered in the chapter. Traffic light icons can be used for your own self-assessment.

Learning checklist

After reading this chapter and completing the activities, I can:

| N3 | L3 | N4 | L4 |

- state that sound waves transfer energy. **Activity 10.1 Experiment 3** ○ ○ ○

- state that humans and animals can hear different frequency ranges. **Activity 10.2** ○ ○ ○

- explain what happens to the pitch and the waveform when the frequency of a sound wave is changed. **Activity 10.3, Activity 10.4** ○ ○ ○

Unit practice assessments

Each unit closes with two sets of practice assessment questions, for N3 and for N4. The questions in these practice assessments help assess the minimum competence for the unit content.

Appendix: Investigating skills

The appendix on pages 366–386 gives guidance on some of the essential scientific skills you will need as you work through this book. You might find it useful to refer to this when carrying out activities.

Answers

Answers to exercises and to the End of Unit assessment questions are available online at:

www.leckieandleckie.co.uk/page/Resources

1 Generation of electricity

Renewable energy sources • Non-renewable energy sources • Generating and distributing electricity • Reducing the impact of energy use

2 Electrical power and efficiency

Electrical power • Calculating power • The cost of electricity • Efficiency

3 Electromagnetism

Magnets and magnetic fields • Electromagnetism • Generating electricity • Practical applications of magnets and electromagnets • Using transformers in high voltage transmission

4 Electrical circuits

Current in a circuit • Voltage in a circuit • Series and parallel circuits • Resistors in circuits • Resistors and Ohm's Law

5 Batteries and cells

Cells and batteries • Fruit cells • Photocells

6 Practical electricity and safety

Electricity in the home

7 Electronic circuits

Electrical devices • Switching devices • Using sensors and logic gates in electronic circuits

8 Gas laws and the kinetic model

Models of matter • Heat transfer • Gas laws and the kinetic model

Unit 1 practice assessment

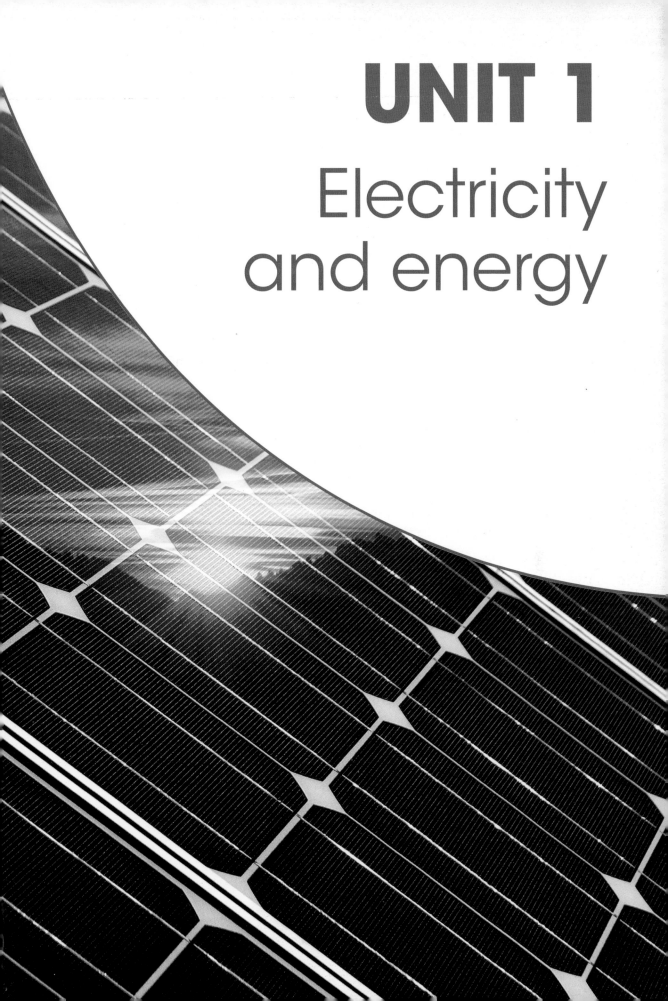

UNIT 1
Electricity and energy

1 Generation of electricity

This chapter includes coverage of:

N3 Energy sources • N4 Generation of electricity
• Planet Earth SCN 3-04b • Planet Earth SCN 4-04a
• Planet Earth SCN 4-04b • Topical science SCN 3-20a
• Topical science SCN 4-20a

You should already know:

- the different types of energy and identify how energy is transferred
- ways of reducing wasted energy
- the difference between renewable and non-renewable sources of energy
- that coal, gas and oil are fossil fuels.

National 3

Curriculum level 3

Planet Earth: Energy sources and sustainability SCN 3-04b

Curriculum level 4

Planet Earth: Energy sources and sustainability SCN 4-04a

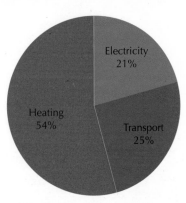

Figure 1.1 *Average energy use in Scotland.*

Renewable energy sources

Learning intentions

In this section you will:

- give examples of renewable energy sources
- describe the advantages and disadvantages of different methods of electricity generation using renewable energy sources.

In daily life, a range of energy sources are used for heating, transport, industrial processes, and electricity generation. Oil and gas are the main sources of energy, accounting for nearly 78% of all energy used in Scotland. Figure 1.1 shows how energy is used in Scotland.

All energy sources can be classified as:

- non-renewable, or
- renewable.

Non-renewable energy sources are fuels such as oil and gas which are destroyed as we use them and are not replaced. Nuclear energy is also a non-renewable energy source. Non-renewable energy sources will run out eventually.

| N3 | L3 | N4 | L4 |

Renewable energy sources are not destroyed when we use them. If we use wind power to generate electricity, we do not reduce the amount of wind that blows or that can be used in the future.

Scotland's energy sources for generating electricity include:

- Gas – this is a non-renewable fossil fuel which will run out at some time in the future. Burning gas produces carbon dioxide, which contributes to global warming. Many countries also use coal for electricity production, but Scotland's last coal-fired power station at Longannet was closed in 2016.

- Nuclear energy – this is a non-renewable form of electricity generation which uses the radioactive material uranium. This will run out at some time in the future. Scotland has two nuclear power stations, at Torness and Hunterston. In 2008, the SNP government voted to oppose the building of any new nuclear power stations in Scotland.

- Hydro – this uses water flow to turn generating turbines. Hydroelectric power is a renewable energy source, but there is a limit to the number of suitable sites for hydroelectric power stations.

- Wind – this uses wind power to turn generating turbines. Wind power is a renewable energy source.

Newer methods of generating electricity use other renewable sources of energy such as biomass, solar, tidal and wave energy. We need to manage our energy demands and develop alternative methods for generating electrical energy, if we are to meet future energy needs.

Scotland's geographic location means there are many places suitable for renewable electricity to be generated. Using renewable energy will reduce greenhouse gas emissions, reduce our consumption of fossil fuels and could create jobs in new technologies. However, many of the best geographical locations in the Highlands and Islands are a long way from the large centres of population in the Central Belt where most electricity is needed. A network of cables is needed to take electricity from where it is generated to where it is required.

Hydroelectric power

Hydroelectric power stations use the kinetic energy of moving water to generate electricity. The gravitational potential energy of water stored behind a dam is transferred to kinetic energy as it runs down pipes. This is used to turn a turbine and a generator and so generate electricity.

📖 **Word bank**

- **Non-renewable energy**
Non-renewable energy comes from sources that will eventually run out.
- **Renewable energy**
Renewable energy comes from sources that will not run out.

❓ **Did you know ...?**
The Scottish Renewable Energy Target aims for 100% of the total Scottish electricity requirement to be generated using renewables by 2020.

Energy change

gravitational potential energy
→ kinetic energy
→ electrical energy

Figure 1.2 *Hydroelectric power station on the shores of Loch Lomond.*

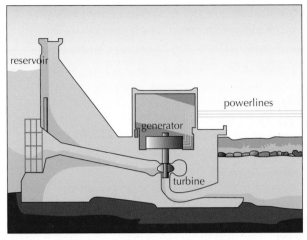

Figure 1.3 *Inside a hydroelectric power station.*

Hydroelectric energy is generally reliable. Hydroelectric power stations usually have a very long working life. Hydroelectric energy provides around 12% of Scotland's total electricity.

Hydroelectric power stations can only be built in a limited number of suitable places and are expensive to build. Most of the hydroelectric schemes in Scotland are in the Highlands.

Hydroelectric power stations can also be used to store energy. Water is pumped up to a storage reservoir, and kept until it is needed. At times of high demand, the water is released down pipes to the generating turbines. These are pumped storage hydroelectric power stations. Scotland's biggest pumped storage power station is at Cruachan, near Loch Awe.

Run-of-river systems are another type of hydroelectric power station. Running water in a river is fed into a generating house that holds a turbine. The water spins the turbine and drives the generator. The used water is fed back into the river. The Atholl Estate in Perthshire operates two run-of-river hydroelectricity schemes. Run-of-river schemes are smaller than conventional hydroelectric power stations and are suitable for local electricity supplies. They can be built in more locations. The energy produced depends on the seasonal flow of the river, making them less reliable than reservoir systems.

Table 1.1 *Different types of hydroelectric power schemes.*

Type of hydro scheme	Advantages	Disadvantages
Hydroelectric generator	Very long life-time.	Can only be built in certain locations.
Generator with pumped storage	Can store energy.	Two reservoirs are needed.
Run-of-river	Can be built in many places.	Depends on flow of the river.

Most hydroelectric power stations in Scotland generate a few tens of megawatts of electricity, although a small number of large stations generate hundreds of megawatts. Run-of-river schemes can generate up to 100 kilowatts. Hydroelectric power stations are usually very efficient, with efficiencies of over 85%. The biggest hydroelectric power station in the world is the Three Gorges Dam in China, which generates 22.5 GW – three times more than is needed for the whole of Scotland.

Wind power

Energy can be generated by harnessing the movement of wind to turn turbines. Wind turbines can be installed in many locations. Small wind turbines can be used to generate electricity for homes and schools. Larger turbines can be used on large agricultural farms or for factories. The turbine blades can vary in size from a few metres for small wind turbines up to nearly 100 metres. The output of the turbine depends on the size of the turbine blades and the strength of the wind.

Wind turbines use electronic sensors to drive motors to make sure the blades are always facing into the wind. This is needed to make sure that the maximum amount of energy is taken from the wind.

Make the link

Power is the energy transferred per second and is measured in watts. Find out more about power in Chapter 2.

★ You need to know

1 megawatt is the same as 1 000 000 watts. 1 GW is 1 gigawatt, which is one thousand megawatts.

Energy change

kinetic energy
→ electrical energy

Figure 1.4   *Small wind turbine for a home.*

Large groups of wind turbines placed together are known as **wind farms**. Scotland has good weather conditions for wind farms. However, the wind does not blow steadily, so wind turbines cannot be guaranteed to generate all the electricity we need when we need it. Also, if the wind is too strong it can damage a turbine. Alternative types of generation are needed to make sure we have a constant supply of electricity.

Scotland's geography means it has many good locations for wind farms, particularly in hilly areas in the Highlands and in the Borders. Wind turbines cause little environmental pollution compared with conventional power stations and can supply electricity in remote locations. However, there are concerns

Figure 1.5   *Whitelee wind farm, near Glasgow.*

about the visual impact of large numbers of wind turbines in remote areas. Some people think wind turbines are an eyesore and should not be built in areas of natural beauty.

Wind farms are also built in offshore locations. The wind is generally stronger and more reliable at sea than it is on land, so offshore wind farms can produce a more reliable supply of electricity. Offshore wind farms are expensive to build.

GO! Activity 1.1   Answers to all activity and assessment questions in this Unit are available online at **www.leckieandleckie.co.uk/page/Resources**

1. Experiment In this activity you will carry out an experiment to show that changing the angle of the blades on a wind turbine affects the generation of electricity.

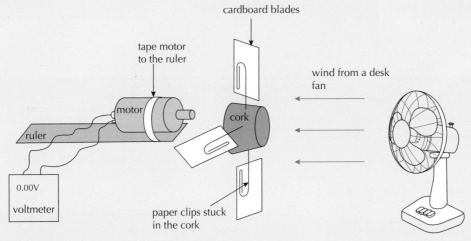

In this model the motor converts the kinetic energy into electrical energy, like a generator. This is the opposite to the usual function of a motor, which converts electrical energy to kinetic energy to make things move.

Read the plan carefully then follow the instructions, making sure you follow all of the safety rules as you would normally be expected to do when carrying out practical work.

You can work in a small group but you should record your own observations and complete the answers yourself.

You will need: small motor, voltmeter, cork, paper clips, cardboard, desk fan, tape, a protractor and a ruler.

Method

(a) Set up your wind turbine as shown in the diagram. Push the cork onto the motor spindle, and mount the cardboard blades on the cork. Make sure everything turns smoothly.

(b) Point the desk fan directly at the wind turbine.

(c) Turn the cardboard blades to be angled at 45° to the direction of the wind from the fan.

N3   L3   N4   L4

(d) Switch on the desk fan and record the voltage. Record your results and repeat your measurements. You could use a table like the one shown below to record your results.

Angle of blades (°)	Voltage (V)			Average voltage (V)
	test 1	test 2	test 3	

(e) Change the angle of the blades. Use the protractor to measure the angle. Record the angle and repeat the experiment.

Repeat this step with the blades at a different angle.

(f) From your observations, write a conclusion about the effect of changing the angle of the blades on the amount of electricity generated.

(g) What variables did you have to control in this experiment?

(h) What could you have done to improve the experiment to get more accurate results?

2. How do you think wind turbines deal with different strengths and different directions of wind?

> **Make the link**
>
> Read about independent, dependent and control variables in the Investigation Skills Appendix.

Biomass

When trees and straw are grown to be used as sources of energy, they are known as biomass crops. Biomass can be used for heating and for transport. In fact, burning biomass is probably the oldest form of energy use on the planet! Burning wood, straw and peat has been used to heat homes for thousands of years. Biomass power stations can also use manure, sewage and even rubbish. This waste produces methane gas, which can be used in a thermal power station.

> **Energy change**
>
> chemical energy
> → heat energy
> → kinetic energy
> → electrical energy.

Energy crops can also be burned in boilers and furnaces to provide direct heating. New technologies can improve the ways in which the heat energy is used, to reduce waste and to provide a consistent source of heat. The Steven's Croft wood-fuelled biomass power station near Lockerbie generates enough electricity to power 70 000 homes every year.

Burning biomass fuels releases carbon dioxide into the atmosphere, although they only release the carbon dioxide that was extracted from the atmosphere during the growing process of the plants. This makes biomass a **carbon neutral** energy source. Being carbon neutral means that the carbon dioxide released by burning that fuel is balanced by locking up the same amount of carbon dioxide in a different part of the process.

> **Word bank**
>
> • **Carbon neutral**
> Carbon neutral energy sources are energy sources or systems which result in no overall carbon dioxide emissions.

The benefits of biomass energy are still debated. Growing the crops may compete with food production. The availability of food may be adversely affected. The production of bioenergy could also lead to deforestation – a change of land use. This may have a negative impact on greenhouse gas emissions. Trees lock up carbon dioxide during the process of photosynthesis. Fewer trees will have a negative impact on the levels of greenhouse gases in the atmosphere.

One way to become more carbon neutral is to buy carbon credits. When companies buy credits they finance renewable energy and forestry projects which help reduce greenhouse gas emissions.

GO! Activity 1.2

😊😊 Do you think any renewable energy resource is carbon neutral? Was any energy used in building the resource? How were the materials brought to the site? Search online to find out more information and discuss your findings in pairs.

GO! Activity 1.3

😊 How do you feel about poo power? Search online to find out more.

Energy change

light energy
→ electrical energy

📖 Word bank

- **Photovoltaic cells**

Photovoltaic cells are used to change light energy into electrical energy.

Make the link

There is more information on photovoltaic cells in Chapter 5.

Solar energy

Solar means 'related to the Sun', so solar energy is the energy we get from the Sun. Solar energy can be used to generate electricity and to provide heating. Solar cells, also known as **photovoltaic cells**, convert sunlight directly into electricity. They are used in a wide range of applications. Small solar cells are used to power watches and calculators, while bigger cells are used to power road signs and timetable displays at bus stops. Large panels can be fitted to roofs to provide electricity for houses, schools, shops and factories.

Solar heating panels absorb heat energy from the sun and transfer this energy to heat water. Individual panels can be installed on houses to provide hot water. Large insulated tanks are needed to make the most of this kind of heating.

Figure 1.6   *Solar panels on a house.*

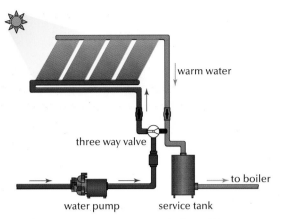

warm water

three way valve

to boiler

water pump   service tank

Figure 1.7   *A thermal solar panel. The cold water flows through the valve to the panel where it is heated. The warm water then flows to the tank to be stored.*

GO! Activity 1.4

☻ A group of pupils investigated the electrical energy produced by a solar cell. They measured the amount of electricity (voltage) produced when they changed the distance between the solar cell and a light source. The graph shows their results.

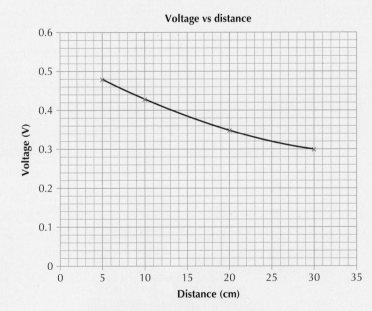

Voltage vs distance

Voltage (V)

Distance (cm)

1. What conclusion can you make from this experiment?
2. What voltage do you think they would have measured at 35 cm from the light source?

(*continued*)

Another group of pupils carried out a similar experiment to measure the effect of placing sheets of tissue paper between the light source and the solar cell. The table shows their results.

Number of layers of tissue paper	Voltage (V)			
	test 1	test 2	test 3	Average
0	0.43	0.52	0.25	
1	0.21	0.22	0.30	
2	0.10	0.10	0.10	
3	0.04	0.03	0.00	
4	0.00	0.00	0.00	
5	0.00	0.00	0.00	

3. Calculate the average voltages.

4. Draw a graph of these results.

5. What real world example do you think the pupils were trying to test and what conclusion could they draw?

If you have access to a small solar cell and a voltmeter you should try this experiment and see if you get similar results. What might be different between these experiments and yours?

Wave energy

Wave power can be used in two different ways:

Energy change

kinetic energy
→ electrical energy

- Shore-based power converters – incoming waves push air up in a column to drive turbines. These systems cause little pollution and do not change the water running through it. They do not generate much electricity for their size.

- Sea-based mechanical devices – linked floats move up and down as waves pass, to drive hydraulic generators.

Figure 1.8 *The Pelamis wave energy converters were longer than a football pitch and were tested off the coast of Orkney.*

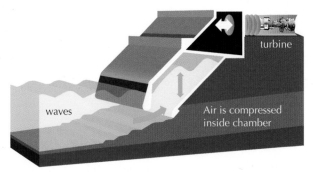

turbine

waves

Air is compressed inside chamber

Figure 1.9 *Inside a shore-based power convertor.*

Tidal energy

Tidal energy plants use the regular motion of tides as they go in and out to turn turbine blades.

Tides are more predictable than wave or wind power. Research is being carried out into generating electricity from massive tidal turbines fixed to the sea bed. Areas of the Scottish coast have high tidal currents due to the topography of the land. These include areas such as between Orkney and Shetland, the Pentland Firth, off the Mull of Kintyre, and around the Hebrides. In the Pentland Firth, the water can be as fast as 3.5–4.5 m/s (7.8–10 miles per hour). This makes it one of the best sites in the world for tidal power.

Energy change

kinetic energy
→ electrical energy.

Geothermal energy

Geothermal power uses the energy that comes from the centre of the Earth. A few kilometres down from the Earth's surface, the temperature can be as high as 250 °C. This heat can be used to turn water to steam, driving turbines and generating electricity. Geothermal energy is a renewable source of energy but can only be built in certain locations. The UK is being surveyed for potential sites. Geothermal energy can also be used for direct heating.

Energy change

heat energy
→ kinetic energy
→ electrical energy.

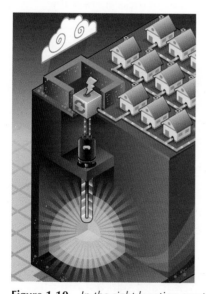

Figure 1.10 *In the right location, geothermal energy can provide electrical energy for homes and schools.*

Advantages and disadvantages of different renewable energy sources for electricity generation

Every method of generating electricity has advantages and disadvantages. Table 1.2 shows some of the advantages and disadvantages of using renewable energy sources for electricity generation.

Table 1.2 *Advantages and disadvantages of renewable energy sources.*

Source of energy	Advantages	Disadvantages
Wind energy	Wind energy is freely available and won't run out.	Only works well in windy places. Many generators are needed to provide electricity for a town.
Hydroelectric energy	Every time it rains, there is more water to provide electricity.	Only suitable in wet, hilly areas. Building dams may damage the environment.
Biofuel (biomass)	Plants are renewable – you can grow some more. It uses natural waste products.	Burning fuel can contribute to air pollution. Growing crops takes up agricultural land which might be needed for food production.
Solar energy	The Sun's energy is freely available whenever the Sun is shining.	Only works well when the Sun shines. Solar cells do not produce much electricity compared to thermal power stations. May need a large area, and they are still quite expensive.
Tidal energy	Tidal energy is freely available and predictable.	Only works well in places where there are strong tides.
Wave energy	Wave energy is freely available and won't run out.	Many kilometres of floats are needed to provide enough electricity for a town.
Geothermal energy	The energy is freely available day or night.	Only possible in parts of the world where the hot rocks are near the surface.

⦿ Activity 1.5

☺ Copy and complete the following passage using the words provided to describe different types of renewable and non-renewable energy sources.

hot water	**non-renewable**	**Sun**	**wind turbines**	**electrical**
geothermal	**wave**	**biomass**	**kinetic**	**renewable**
hydroelectric	**advantage**	**disadvantage**		

_ _ _ _ _ _ _ _ _ energy comes from sources that will not run out. _ _ _ _ _ _ _ _ _ _ _ _ _ _ energy comes from finite sources; once they have been used they are no longer available to generate more energy.

Solar power uses energy from the _ _ _ and can be used to generate electricity and _ _ _ _ _ _ _ _ _. An _ _ _ _ _ _ _ _ _ of solar energy is that it will not run out. A _ _ _ _ _ _ _ _ _ _ _ _ is that a lot of solar panels are needed to produce lots of electricity.

Water can be used to generate electricity in _ _ _ _ _ _ _ _ _ _ _ _ _ power stations.

_ _ _ _ _ _ _ _ _ _ _ _ transfer the _ _ _ _ _ _ _ energy of the wind into _ _ _ _ _ _ _ _ _ _ energy.

Other examples of renewable energy sources include _ _ _ _ energy, _ _ _ _ _ _ _ energy and _ _ _ _ _ _ _ _ _ _.

N3 L3 N4 L4

Non-renewable energy sources

National 3

National 4

Curriculum level 4

Planet Earth: Energy sources and sustainability
SCN 4-04a, 4-04b

Learning intentions

In this section you will:

- give examples of non-renewable energy sources
- describe the formation of fossil fuels
- describe the advantages and disadvantages of different methods of electrical generation using non-renewable energy sources.

Most of our energy needs in the United Kingdom are met using non-renewable fossil fuels – coal, oil and gas. As the fossil fuels are running out it is important that we conserve the supply we have. When fossil fuels are burned, they emit carbon dioxide (CO_2). This is a greenhouse gas. Increasing levels of greenhouse gases in the atmosphere are contributing to global warming. There is international agreement that we need to reduce the use of fossil fuels to reduce carbon dioxide emissions. The UK also uses nuclear fuelled power stations. Nuclear power is non-renewable, but does not emit CO_2, so some people see it as an answer to the problem of global warming.

Fossil fuels

Coal is formed from dead plants which became buried under layers of mud. The mud stops the plant matter from rotting away. Heat and pressure cause the plants to turn into coal. This takes millions of years.

Oil and gas are formed in a similar way. They were originally tiny plants and animals that lived in the sea. Their remains turned to oil and gas as they experienced heat and pressure. The oil and gas often get pushed back up due to underground pressures to form reservoirs under a layer of rock.

Conventional thermal power stations use coal and gas as their fuel. These power stations are cheap to run and can produce a constant supply of energy. They are not dependent on the weather. Coal- and gas-fired power stations use coal and gas to heat water to very high temperatures to make steam in a high-pressure vessel. The high-pressure steam is used to drive turbines, which turn the generators.

Coal and gas are convenient fuels. They are easy to process and it is easy to control the amounts of electricity produced.

★ You need to know

Global warming describes the increase in the temperature of the Earth's oceans and atmosphere arising from the increases in amounts of carbon dioxide and other greenhouse gases in the atmosphere.

Figure 1.11 *Coal, oil and gas.*

Energy change

chemical energy
→ heat energy
→ kinetic energy
→ electrical energy.

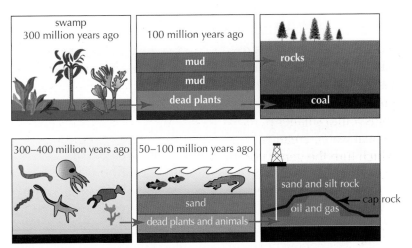

Figure 1.12 *Coal, oil and natural gas were formed over millions of years.*

Two problems with this way of generating electricity are:

- the fuel will run out eventually

- the raw materials are needed in other industries.

As well as being used for electricity generation, fossil fuels are used for:

- Transport – most cars and lorries and other road vehicles use petrol or diesel; planes use gasoline; some trains use diesel; ships use diesel fuels.

- Direct heating – domestic and commercial heating often uses gas or oil, and coal is used in industrial heating.

- Industrial processes – oil is used for the manufacture of plastics and paints. Many clothes are made of synthetic materials which are produced using oil.

Most manufacturing processes, such as steel production, paper milling, chemical manufacturing and engineering use fossil fuels for heating and for raw materials.

↻ Keep up to date!

Search online to find out how long we can rely on coal, oil and gas to meet our energy needs.

⏵ Activity 1.6

☺ Answer true (T) or false (F) to these questions.

1. Fossil fuels are formed from prehistoric plants and animals.
2. Power stations can use fossil fuels in any weather.
3. Nuclear energy is a type of fossil fuel.
4. Carbon dioxide is given off when fossil fuels are burned.
5. We can use fossil fuels as much as we like, because they will never run out.
6. Thermal power stations are an example of non-renewable energy sources.

Nuclear energy

Nuclear power stations use radioactive material to generate electricity. The radioactive material produces heat energy inside a reactor, in a process called **fission**. The energy is produced when a neutron hits the nucleus of an atom and splits it into smaller parts, as shown in Figure 1.13. This is used to heat water, creating steam. The steam drives the turbines which turn the generator.

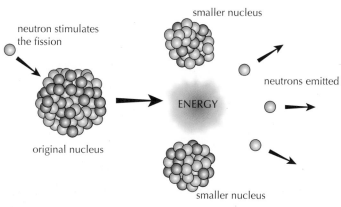

neutron stimulates the fission

smaller nucleus

neutrons emitted

ENERGY

original nucleus

smaller nucleus

Figure 1.13 *Nuclear fission.*

There are advantages to using nuclear power to produce energy. Less nuclear fuel is required to generate the same amount of electricity than using a fossil fuel. 1 kg of nuclear fuel can provide two to three million times more energy than the same mass of a fossil fuel. No carbon dioxide is released from nuclear power stations.

However, nuclear power plants are expensive to build, and the costs of the disposal of radioactive waste and dismantling the reactor safely can be very high.

When nuclear fuel has undergone the fission processes needed to generate electricity, it becomes **nuclear waste**. Nuclear waste is radioactive and so it is a danger to human health and to the environment. It must be securely stored for many years to prevent nuclear accidents. The significant risk to human health and the environment has caused people to question whether nuclear power is a suitable alternative to fossil fuels.

Figure 1.14 *Torness nuclear power station in East Lothian.*

📖 Word bank

• **Nuclear waste**

Waste that contains radioactive materials, produced in nuclear power stations and in industrial and medical applications.

Make the link

Find out more about nuclear reactions and nuclear waste in Chapter 13.

Activity 1.7

☺ Use this activity to review what you have learned about nuclear power.

1. What type of energy is stored in radioactive material?
2. Give an advantage and disadvantage of nuclear power.
3. Research, plan and carry out a debate on nuclear fuel. Should more nuclear power stations be built, and if so, where?

Advantages and disadvantages of different non-renewable energy sources for electricity generation

Non-renewable energy sources generate electricity using resources which cannot be naturally replenished in a human timescale. Once a fuel source such as gas has been used, we cannot get it back again.

There is one major gas-fired power station in Scotland, at Peterhead, and two nuclear power stations. The table shows the advantages and disadvantages of non-renewable energy sources for electricity generation.

Table 1.3 *Advantages and disadvantages of non-renewable energy sources.*

Source of energy	Advantages	Disadvantages
Fossil fuels	Not weather-dependent, easy to process, easy to store.	Emission of CO_2 contributes to global warming. Fuel will run out.
Nuclear fuel	Provides constant level of reliable generation.	Nuclear waste has to be safely stored. Expensive and potentially dangerous.

National 4

Generator

Turbine

blades

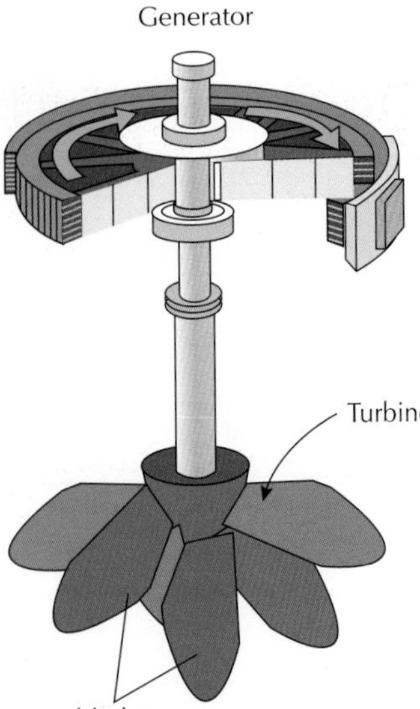

Figure 1.15 *A turbine generator.*

📖 Word bank

• **Turbine**

A turbine is a machine that is made to rotate by a fast-flowing stream of water or gas. It is often used to turn parts of a generator.

Generating and distributing electricity

Learning intentions

In this section you will:
• explain the basic function of a thermal power station
• identify the issues of energy efficiency related to electricity distribution.

In a thermal power station, heat energy turns water to pressurised steam which turns the blades of a **turbine**. The rotating turbine is connected to a generator. As the electromagnets in the generator are turned it causes electrical charges to flow through the wires.

The heat energy comes from burning fossil fuels such as coal or gas or from a nuclear reactor.

Most old thermal power stations which use coal or gas are not very **efficient**. Only about 40% of the energy in the fuel can be transferred to electricity, with 60% of the energy being wasted as heat energy, which is transferred to the surroundings through the cooling towers and chimneys of the power station. When steam leaves the turbines, it has to be cooled before being heated again in the boiler, wasting more energy.

| N3 | L3 | **N4** | L4 |

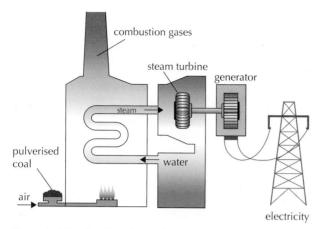

Figure 1.16 *Inside a thermal power station.*

Labels in figure: combustion gases, steam turbine, generator, steam, pulverised coal, water, air, electricity

Newer power stations which use gas are much more efficient. The gas-fired power station at Peterhead is about 57% efficient, which means that only about 43% of the energy in the fuel is wasted.

Combined heat and power stations (CHP) provide heat and electricity, by using some of the heat energy for heating. The energy is usually used to heat water, but can also be used to heat industrial processes in chemical factories. In towns, the hot water can be used to heat houses and schools near the power station. This can use up to 75% of the original chemical energy stored in the fossil fuel. The INEOS oil refinery at Grangemouth uses CHP to provide electricity and heat for the whole industrial site.

Distribution of electrical energy

Electricity is distributed across the United Kingdom using the National Grid. This is a national distribution network of cables that carries electrical energy from power stations around the country to where it is needed. Some energy is wasted in the distribution network because of the heating that happens when electrical charges pass through the cables.

To reduce this wasted energy, **transformers** are used to increase the voltage of the electrical energy near the power station. A step-up transformer is used to increase the voltage from around 25 000 volts to 400 000 volts before the energy travels along low-resistance wires. At the other end, in local substations, step-down transformers reduce the voltage to 230 volts so that the electrical energy is safe to be distributed to homes.

About 2.3% of the electrical energy generated in power stations is wasted in the National Grid.

Figure 1.17 *Power substation.*

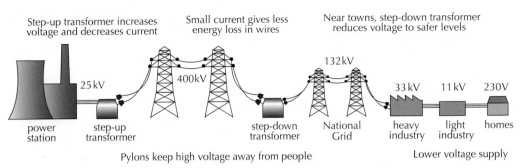

Step-up transformer increases voltage and decreases current

Small current gives less energy loss in wires

Near towns, step-down transformer reduces voltage to safer levels

132kV

25 kV   400kV   33 kV   11 kV   230V

power station   step-up transformer   step-down transformer   National Grid   heavy industry   light industry   homes

Pylons keep high voltage away from people

Lower voltage supply

Figure 1.18   *Electrical energy is transferred from power stations to homes using transformers and power lines.*

 Activity 1.8

In this activity, your teacher will demonstrate:
- the loss of electrical energy when power lines are used
- the use of transformers to reduce the amount of energy wasted in power lines.

⚠ The experiment has high voltages and should only be seen as a demonstration by a qualified science teacher.

In Figures 1.19 and 1.20, the a.c. supply represents the power station. The first bulb represents a home near the power station and the second bulb represents a home further away that needs the power lines to transfer the energy.

In Figure 1.19, some energy is wasted as heat in the power lines. Not all the electrical energy is transferred through the power lines, so the second bulb is not as bright as the first. The voltage drops from 12 V to 3 V. The difference in energy is due to heat energy being wasted in the wires.

In Figure 1.20, transformers are added to represent the substations. The voltage is stepped up for transmission along the power lines to reduce energy being wasted. It is then stepped down again by the second transformer. The second bulb is as bright as the first.

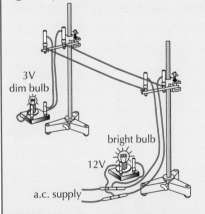

3V dim bulb

bright bulb

12V

a.c. supply

Figure 1.19   *Power lines demonstration where electrical energy is wasted.*

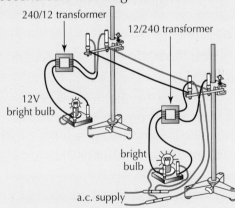

240/12 transformer

12/240 transformer

12V bright bulb

bright bulb

a.c. supply

Figure 1.20   *Transformers being used to reduce the energy being wasted.*

Activity 1.9

☺ **1.** Describe the basic way a thermal power station works to generate electricity.

2. How much energy is wasted in

(a) an old thermal power station   (b) a CHP station   (c) the National Grid?

3. What type of transformers are used to increase the voltage in distribution networks?

| N3 | L3 | N4 | L4 |

Reducing the impact of energy use

Curriculum level 3

Topical Science SCN 3-20a

National 4

Curriculum level 4

Topical Science SCN 4-20a

Learning intentions

In this section you will:

- give an example of new technologies being developed to reduce the demand on fossil fuels
- give examples of ways to reduce energy use.

All energy use and electricity production has an impact on the world and the environment. To reduce the harmful impacts, we need to:

- reduce the impact from existing forms of energy uses
- find new technologies which have less impact.

We also need to make the best use of the energy sources we already have, such as fossil fuels and nuclear power. These sources will eventually run out. New technologies must be developed to meet the increasing demand for electrical energy and to reduce the use of fossil fuels. Scientists are researching and experimenting with different ideas.

Energy for transport

Most transport uses fuels such as petrol and diesel, which are forms of oil. Transport accounts for about a quarter of all Scotland's energy use. Petrol and diesel emit carbon dioxide when they are burned in engines, so they contribute to global warming. They also emit other polluting gases and particles, which lead to poor air quality. It has been estimated that air pollution in Scotland causes over 2000 deaths each year. Alternative types of transport use different types of fuel to reduce the polluting impact of conventional petrol and diesel engines.

Hybrid and electric cars

Hybrid engines use combinations of electric motors and conventional engines. In some hybrids, the kinetic energy of the car is used to generate electricity, which can be stored in rechargeable batteries. This stored energy can then be used to drive the motor. Other types of hybrid vehicles are plugged into charging points to charge the batteries.

Electric vehicles use batteries to drive electric motors. Most electric vehicles need to be charged at charging points, which can be hard to find in remote locations. Currently, electric vehicles do not have the same range as conventional vehicles,

Figure 1.21 *Electric car charging at a city-centre charging point.*

so they are less suitable for long journeys. However, they do not pollute the atmosphere and they are very suitable for short journeys in towns and cities.

Biofuels and biodiesel

Biofuels are fuels made from biological processes, using agricultural crops and industrial and commercial waste. Waste fats, oils and greases are collected from industry and retailers such as fish and chip shops and processed into biodiesel for use in diesel engines in cars and lorries. Diesel fuel sold in the UK is a mixture of conventional diesel fuel and biodiesel. A distillation plant in Motherwell supplies over 4 million litres of biodiesel throughout the UK each year.

Hydrogen fuel

Liquefied hydrogen gas can be used as a fuel. The stored chemical energy in hydrogen is converted to kinetic energy, either by burning the fuel in an internal combustion engine or by chemically reacting hydrogen with oxygen in a fuel cell to run electric motors. Aberdeen was the first European city to offer hydrogen fuelled cars as part of a car-share club. Aberdeen also has a fleet of hydrogen fuelled buses. These buses do not use fossil fuels, so they help reduce pollution and conserve supplies of fossil fuels.

In Unst, the most northerly inhabited island of Scotland, electricity produced by renewable energy systems is being stored as hydrogen. A process called electrolysis is used to make hydrogen from water. This can then be used as fuel for cars, vans, buses or even ferries.

Figure 1.22 *A toy hydrogen car. A solar cell on a fuel tank splits the water into hydrogen and oxygen. The hydrogen is then pumped into the car and converted to electrical energy, which can make the car move.*

Energy supply and demand

The demand for electricity varies throughout the year and throughout each day. Demand is also affected by irregular events, such as particularly extreme weather conditions or very popular TV programmes. On a cold winter evening, the demand for electricity can be double the demand on a summer day. To make sure we always have enough electricity to meet demand, different types of electricity supply are used. Pumped storage hydroelectric schemes can be used to meet sudden surges in demand.

Although renewable electricity from wind farms reduces our use of fossil fuels, it is not as reliable. The energy available from a wind farm is affected by the strength of the wind, and if there is no wind, there is no electricity. Backup generators are needed to make sure there is no drop in supply.

To help match electricity supply and demand, it is also possible to reduce demand.

N3   L3   N4   L4

Changing the way you live could have a great effect on the amount of energy you use. You could:

- take a bus instead of using a car

- wear a jumper instead of turning the heating up

- boil just the water you need instead of filling the kettle to the top

- make sure your home is well-insulated and draught-proofed so it doesn't lose heat to the environment

- switch off appliances and lights when they are not needed

- use energy-efficient appliances.

Using energy more efficiently helps to:

- conserve fossil fuel supplies

- reduce any harmful environmental impacts of generating energy

- save money.

Big industrial users of electricity such as factories and chemical plants can be asked to change their manufacturing processes so their demand is at a different time to domestic demands.

The UK Government plans for every home and business to have a smart meter for electricity and gas by the end of 2020. Smart meters will keep track of the energy you use in your home and help you use it more efficiently.

❓ Did you know ...?

The Scottish Government wants to make sure that everyone in Scotland who needs help with their energy costs gets support. The Home Energy Efficiency Programmes for Scotland (HEEPS) makes it easier to get an efficient new boiler. A more efficient boiler transfers more energy to heating the water and saves energy and money.

GO! Activity 1.10

☺ Create a poster encouraging others to save energy. Search online to help you find facts for your work. Don't forget to reference the website you use.

Make the link

How to reference a website is described in the Investigation Skills section in the Appendix.

GO! Activity 1.11

☺ What would you say if another pupil said this to you?

Find out what your school is doing to reduce its energy use.

Do you have an eco-group? If not, have you thought about starting one?

I don't care about how much energy I use. It's not my problem if the fossil fuels run out.

Two pupils discussing energy.

Cleaner energy

Carbon capture is a process designed to remove carbon dioxide out of the waste gases produced by fossil fuel power stations. The carbon dioxide can be transported and stored in deep underground geologic formations. Scotland is well-placed to use carbon capture and storage, with a number of potential storage sites in the North Sea, but the technology required makes it an expensive process.

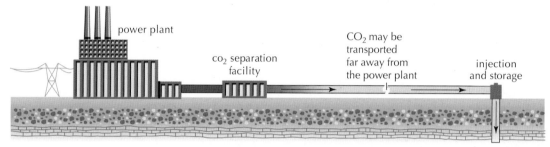

power plant

CO_2 separation facility

CO_2 may be transported far away from the power plant

injection and storage

Figure 1.23   *The carbon capture and storage process.*

Nuclear fusion

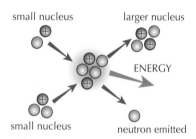

small nucleus

larger nucleus

ENERGY

small nucleus

neutron emitted

Figure 1.24   *Nuclear fusion.*

Nuclear fusion is an alternative type of nuclear energy. Nuclear fusion occurs when two smaller **nuclei** join to form a new, larger particle. When this happens, large amounts of energy are released. This is the process that is happening all the time in the Sun. Scientists are working to make nuclear fusion power stations a possible option for our future energy needs.

Experimental fusion reactors are being developed in France, Germany, the USA and Japan.

Solar updraft towers

A solar updraft tower is a renewable energy power station that uses low-temperature heat energy from the Sun to generate electricity. The movement of hot air up the chimney drives turbines which produce electricity. Small amounts of heat energy are needed to generate electricity due to the large area at the base of the tower.

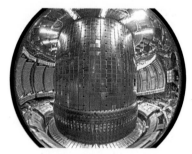

Figure 1.25   *Inside the nuclear fusion reactor, Tokamak.*

📖 Word bank

- **Nucleus and nuclei**

The nucleus is the centre of an atom. It is positively charged and contains nearly all the mass of the atom. The plural of 'nucleus' is 'nuclei'.

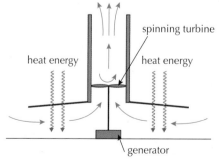

spinning turbine

heat energy

heat energy

generator

Figure 1.26   *A solar updraft tower. The heat energy causes the movement of air past the turbine.*

N3   L3   N4   L4

GO! Activity 1.12

1. Experiment In this activity you will make a model of a solar updraft tower.

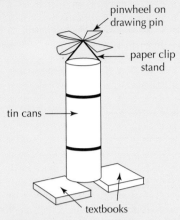

A model solar updraft tower.

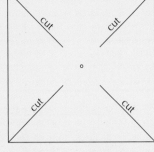

Template for a pinwheel.

Read the plan carefully then follow the instructions, making sure you follow all of the safety rules as you would normally be expected to do when carrying out practical work.

You can work in a small group but you should record your own observations and complete the answers yourself.

You will need: three empty, large, clean tin cans with the tops and bottoms removed, a paper clip and drawing pins, stiff card, sticky tape, two blocks or textbooks.

Method

   (a) Tape together the cans to form a long tube.
   (b) Unbend the paper clip and use the wire to form an arch across the top of the cans, sticking it in place with some tape.
   (c) Attach a drawing pin, point up, to the top of the wire arch with some tape.
   (d) Using a square piece of paper and copying the template given here, make a pinwheel. Cut diagonally in from the corners, leaving a gap in the middle. Fold over every other point to the centre and tape them in place.
   (e) Balance the pinwheel on top of the drawing pin.
   (f) Rest your solar updraft tower on two books and place it in the sunshine.
   (g) Why does the pinwheel turn?
   (h) What would happen if the tower was not on the books?
   (i) What if the tower was painted black?
   (j) Could enough power be generated by the spinning pinwheel to turn a generator?
   2. Why is it important to develop renewable energy resources that work in different locations around the world, where levels of sunshine might be different?
   3. Search online to research solar updraft towers.

> ### 🔍 Hint
> Look at the information on radiation in Chapter 8.

GO! Activity 1.13

☺ Different energy sources can be used for different purposes.

Complete the table by ticking the boxes to show which energy sources can be used for generating electricity, heating and transport. Some entries have already been done for you.

	Electricity	Heat	Transport
Oil	✓	✓	✓
Coal			
Gas			
Wind			
Waves			
Tides	✓		
Biomass			
Hydrogen			
Solar			
Geothermal	✓	✓	
Nuclear fusion			

Learning checklist

After reading this chapter and completing the activities, I can:

N3 L3 N4 L4

• state that three examples of renewable energy sources are solar energy, wind energy and hydroelectric energy. **Activity 1.1, Activity 1.2, Activity 1.4**	○ ○ ○
• describe the advantages and disadvantages of different methods of electrical generation using renewable energy sources. **Activity 1.2, Activity 1.5**	○ ○ ○
• state that four examples of non-renewable energy sources are coal, oil, gas and nuclear energy. **Activity 1.6**	○ ○ ○
• describe the formation of fossil fuels. **Activity 1.6**	○ ○ ○
• describe the advantages and disadvantages of different methods of electrical generation using non-renewable energy sources. **Activity 1.6, Activity 1.7**	○ ○ ○

N3 **L3 N4 L4**

- state that transformers are used to improve the efficiency in electricity distribution. **Activity 1.8** ○ ○ ○

- explain the basic function of a thermal power station. **Activity 1.9** ○ ○ ○

- state that two examples of ways to reduce energy use are using energy-efficient appliances or driving a hydrogen fuelled car. **Activity 1.10, Activity 1.11** ○ ○ ○

- state that three examples of new technologies being developed to reduce the demand on fossil fuels are carbon capture, fusion reactions and solar updraft towers. **Activity 1.12** ○ ○ ○

2 Electrical power and efficiency

This chapter includes coverage of:

N3 Electricity • N4 Electrical power

You should already know:

- that electrical devices transfer energy from one form to another
- that electrical energy is generated in power stations or from renewable energy sources.

National 3

National 4

Electrical power

Learning intentions

In this section you will:

- state that power is a measure of the energy transferred every second.
- state that different electrical appliances use different amounts of electrical power.

Energy is needed to make things work. Electrical devices convert electrical energy into other forms of energy, such as sound, heat, light and kinetic energy (movement). For example, an electric kettle changes electrical energy to heat energy and sound energy. The heat energy is the **useful energy** – a kettle should heat up water – and the sound energy is **wasted energy** – the sound produced by a kettle doesn't serve any purpose.

Most electrical systems and devices will transfer energy into useful and wasted energy.

Energy (E) is measured in joules (J). Electrical appliances convert energy at different rates. **Power** (**P**) is a measure of the amount of energy transferred every second. Power is measured in **watts** (**W**).

Appliances such as electric kettles are rated in terms of their power. This is a measure of the rate at which they use energy.

📖 Word bank

- **Power**

Power is the rate of energy transfer, or energy transferred per second.

Electrical power ratings

We can use **power ratings** to compare how much power different appliances use. The table shows typical power ratings for a range of common household appliances.

Table 2.1 *Example power ratings for different appliances.*

Appliance	Power (W)
Oven	3000
Dishwasher	1400
Iron	900
Hairdryer	1600
Microwave	1000
TV	150
Stereo	60
Filament lamp	100
Energy-saving lamp	12
Drill	750

Domestic electrical appliances usually have power rating labels attached to them. The label gives information about the electrical device including the:

- power rating
- voltage
- frequency

at which the appliance is designed to work. The highest power rated devices will usually be those designed to convert electrical energy into heat energy, such as ovens, dishwashers and irons.

Some appliances transfer electrical energy to more than one other form of energy. For example, a hairdryer transfers electrical energy to heat energy and kinetic (movement) energy in order to blow out hot air.

Figure 2.2 shows the power rating label for an electric heater, rated at 2000W. This means the heater transfers 2000 joules of energy every second.

The information on power rating labels can be used to make sure the domestic wiring circuits in your house are suitable for the power requirements of all the appliances. The power rating labels also tell you which appliances will be most expensive to run. Considering the values on the rating labels can help save money, energy and the environment.

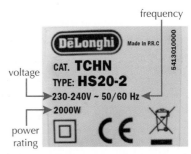

Figure 2.2 *Power rating label from an electric heater.*

GO! Activity 2.1

 1. Write out the useful energy changes for the list of appliances in Table 2.1.

2. Which appliance in Table 2.1 transfers the most energy every second?

3. Explain the difference between energy and power.

4. Which of these two appliances transfers energy faster: a 1600 W hairdryer or a 2000 W fan heater?

5. Look at the information in Table 2.1. Would you save more energy (or money)

  (a) switching off the TV when you leave a room or

  (b) switching off an energy-saving light bulb?

  Explain your answer.

The amount of energy used by different appliances can be monitored by using:

- a plug-in energy monitor
- a joulemeter.

A plug-in energy monitor plugs into the household sockets and is an easy way to measure energy, power or current and voltage of a household appliance. A joulemeter is a piece of scientific apparatus that can be used in a laboratory to measure energy. It is also possible to use the new smart energy meters.

GO! Activity 2.2

1. Experiment In this activity you will use a plug-in energy monitor to measure the amount of energy used by different appliances. The plug-in energy monitor measures energy being transferred every second.

Read the plan carefully then follow the instructions, making sure you follow all of the safety rules as you would normally be expected to do when carrying out practical work.

You can work in a small group but you should complete the answers yourself.

You will need: a plug-in energy monitor and a range of appliances such as a microwave, a hairdryer, a computer, a lab power pack, a television and a mobile phone charger.

A plug-in power and energy monitor.

Method

  (a) Plug in and switch on each appliance.

  (b) Measure the power. Do not leave the appliances running for too long.

  (c) Draw a suitable table and record your results.

  (d) What type of chart or graph could be used to display these results?

Measuring the power of a microwave.

GO! Activity 2.3

☺☺☺ **1. Experiment** In this activity you will use a joulemeter to measure the amount of energy used by different appliances. A joulemeter measures the energy transferred by the component.

Read the plan carefully then follow the instructions, making sure you follow all of the safety rules as you would normally be expected to do when carrying out practical work.

You can work in a small group but you should record your own observations and complete the answers yourself.

You will need: a lab power pack, four connecting leads, a joulemeter, a stopwatch and a mounted lamp/bulb.

Circuit diagram, measuring the energy transferred in a light bulb.

Method

(a) Set up the circuit as shown in the diagram.

(b) Reset the joulemeter to zero.

(c) Switch on the lamp and start the stopwatch at the same time.

(d) Record the numbers shown on the joulemeter after 100 seconds.

(e) Calculate the energy used in 1 second.

(f) Compare this value to the power rating on the bulb. What do you notice? If you have other power rated bulbs available test your conclusion.

Calculating power

National 4

Learning intention

In this section you will:

• use the relationship between power, energy and time.

Power is defined as the **energy transferred per second**, or the rate of energy transfer.

For electrical devices, power is a measure of the electrical energy transferred to an appliance every second. One watt is the same as one joule per second. A 12 W light bulb transfers 12 J of energy every second to light energy and heat energy.

Power, energy and time are linked in the relationship shown here.

> **Power = energy ÷ time.**
> This can be written using the symbols
> $$P = \frac{E}{t}$$

Hint

When solving a problem using a relationship it is important to make sure that the values put into the equation have the correct units. You may have to change time in hours or minutes to time in seconds. Remember there are 3600 seconds in 1 hour.

Table 2.2 *Symbols and units for the power relationship.*

Name	Symbol	Unit	Unit symbol
power	P	watts	W
energy	E	joules	J
time	t	seconds	s

Example 2.1

An electric kettle transforms 180 000 J of electrical energy into heat energy and sound energy in a time of 120 seconds.

Calculate the power of the kettle.

energy = 180 000 J

time = 120 s

power = ? ⟵ Write out what you know from the question.

$\text{power} = \dfrac{\text{energy}}{\text{time}}$ ⟵ Write out the equation.

$\text{power} = \dfrac{180\,000}{120}$ ⟵ Substitute in what you know.

power = 1500 W ⟵ Don't forget to write the units in your answer.

Example 2.2

The power rating of a food mixer is 350 W. It is used for 3 minutes.

Calculate the energy transferred.

$t = 3 \text{ minutes} = 3 \times 60\,\text{s} = 180\,\text{s}$

$P = 350\,\text{W}$

$E = ?$ ⟵ Write out what you know from the question.

$P = \dfrac{E}{t}$ ⟵ Write out the equation.

$350 = \dfrac{E}{180}$ ⟵ Substitute in what you know.

$350 \times 180 = E$

$E = 63\,000\,\text{J}$ ⟵ Don't forget to write the units in your answer.

Example 2.3

A 1000 W toaster transfers 200 000 J when it is toasting two slices of bread.

Calculate the time the toaster was working for.

$P = 1000\,W$

$E = 200\,000\,J$ ———— Write out what you know from the question.

$P = \dfrac{E}{t}$ ———— Write out the equation.

$1000 = \dfrac{200\,000}{t}$ ———— Substitute in what you know.

$t = \dfrac{200\,000}{1000}$

$t = 200\,s$ ———— Don't forget to write the units in your answer.

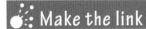

Make the link

There is more information on using relationships in the Investigation Skills Appendix.

GO! Activity 2.4

☺ **1.** Using the power, energy and time relationship, complete this table for missing entries **(a)–(g)**.

Power (W)	Energy transferred (J)	Time (s)
(a)	100	5
(b)	7500	3000
(c)	250 000	2000
12	(d)	300
400	400	(e)
1400	21 000	(f)
24	10 800	(g)

2. A 1400 W heater transfers 168 000 J of energy. Calculate the time that it was switched on for.

3. A 400 W hairdryer was switched on for 40 seconds. Calculate the energy used.

4. A 700 W washing machine was running for 2 hours. Calculate how much energy was used.

(continued)

| N3 | L3 | **N4** | L4 |

5. An electric fire uses 5 220 000 J of electrical energy in 30 minutes. Calculate the power rating of the fire.

6. Carolyn listens to a CD player for 30 minutes. The power rating of the CD player is 15 W. Calculate the energy transferred in this time.

7. Calculate the amount of energy transferred by a set of Christmas lights that have a power rating of 40 W when they are left on for 4 hours.

National 3

The cost of electricity

Learning intention

In this section you will:

• explain how to calculate the cost of electricity in household situations.

We have to pay for the electrical energy that appliances use. The **cost** depends on the **power rating** of the appliance and the **time** that it is used for. To calculate the cost, we need to work out the number of **kilowatt hours** (kWh) used and the time. 1 kilowatt hour is the amount of electrical energy consumed by a 1 kW (1000 W) rated appliance in 1 hour. Kilowatt hours is a measure of electrical energy and is calculated using the relationship shown here.

> **number of kilowatt hours = power × time**
>
> where power is measured in kilowatts and time is measured in hours

The total cost of the electrical energy used is found using the relationship shown here.

> **total cost = number of kilowatt hours × unit cost**
>
> where the unit cost is the cost of 1 kWh.

Example 2.4

A 0.1 kW incandescent light bulb is left on overnight for 8 hours. If the unit cost of electricity is 12 pence what is the cost (to the nearest penny) of leaving the light on?

$P = 0.1$ kW

$t = 8$ hours ●————————————(Write out what you know from the question.)

unit cost = 12 p

number of kilowatt hours = ?

total cost of energy used = ?

number of kilowatt hours = $P \times t$ ●————————(Work out the number of kWh.)

number of kilowatt hours = 0.1×8 ●————————(Substitute in what you know.)

number of kilowatt hours = 0.8 kWh

total cost = number of kilowatt hours × unit cost ●———(Work out the total cost of the energy used.)

total cost = 0.8×12

total cost = 9.6 p, or 10 p to the nearest penny.

GO! Activity 2.5

☺ **1.** A 2.4 kW kettle is used during the day for a total of 1 hour. The unit cost of electricity is 7 p. Calculate the cost (to the nearest penny) of using the kettle.

2. A 0.15 kW TV is used to watch a film of 2 hours in length. The unit cost of electricity is 9 p. Calculate the cost (to the nearest penny) of watching the film.

3. A 1.2 kW iron is used for 1 hour. The unit cost of electricity is 8 p. Calculate the cost (to the nearest penny) of using the iron.

4. A 0.04 kW set of Christmas lights is left on all night, for 8 hours. The unit cost of electricity during the night is 4 p. Calculate the cost (to the nearest penny) of leaving the lights on.

5. Anna does a load of washing at lunchtime and another before she goes to bed at night. Each cycle takes 2 hours.

The unit cost of electricity during the day is 7 p.

The unit cost of electricity at night is 3 p.

Calculate the cost (to the nearest penny) of using the 0.5 kW washing machine for both loads. What is the difference in price?

Reading an electricity bill

Electricity bills show meter readings at the start and end of a billing period. The difference between the readings is the number of units of electrical energy used in that billing period.

The total cost of electricity used is calculated by multiplying the number of units used by the unit cost. Many energy suppliers also have a standing charge, to cover the cost of maintaining the distribution network. The unit cost will be slightly different for different suppliers.

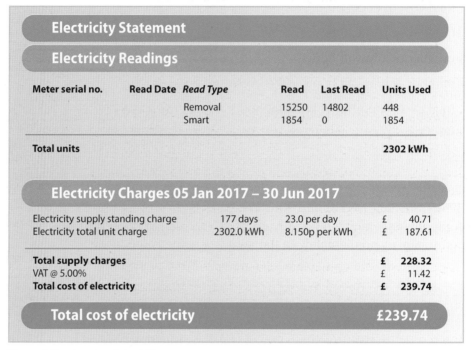

Electricity Statement

Electricity Readings

Meter serial no.	Read Date	Read Type	Read	Last Read	Units Used
		Removal	15250	14802	448
		Smart	1854	0	1854
Total units					**2302 kWh**

Electricity Charges 05 Jan 2017 – 30 Jun 2017

Electricity supply standing charge	177 days	23.0 per day	£	40.71
Electricity total unit charge	2302.0 kWh	8.150p per kWh	£	187.61
Total supply charges			**£**	**228.32**
VAT @ 5.00%			£	11.42
Total cost of electricity			**£**	**239.74**

Total cost of electricity	**£239.74**

Figure 2.3 *An electricity bill.*

GO! Activity 2.6

☻ Search online find out about different suppliers and costs of domestic electricity.

National 3

National 4

Efficiency

Learning intentions

In this section you will:

- describe how being energy-efficient can conserve resources and the environment
- use the relationship between input and output to calculate efficiency using power and energy.

The **efficiency** of a process or system is a measure of the useful output energy or power compared with the input energy or power. Efficiency can be a measure of how well something works, or how well the process or system uses the energy.

An efficient process or system is one in which as much input energy as possible is converted into useful energy, without wasting too much energy. Efficiency is usually expressed as a percentage. Efficiency is always less than 100%, because it is impossible to get the same or more energy out than was put in. Efficiency applies to mechanical and electrical devices.

> 📖 **Word bank**
>
> • **Efficiency**
> Efficiency is a measure of how much of the total input is transferred to useful output.

Wasted energy

Some appliances are designed to transfer electrical energy to heat energy, such as kettles, toasters and hair straighteners. Other appliances such as TVs use electricity to produce sound and light, but they also produce heat – your TV will warm up after it has been switched on for a while. This heat energy is called **wasted energy**.

Some appliances need to be cooled to stop them overheating. For example, many desktop and laptop computers get warm during use and need to have fans in the casing to prevent the components getting too hot. Overheating can lead to poor performance and ultimately to the failure of some components.

Figure 2.4 *Fans inside a computer case stop the computer overheating.*

Domestic appliances may have an energy efficiency label. These were introduced in the 1990s and tend to be on larger appliances such as washing machines and fridges. Energy efficiency labels help customers to compare products easily. The most efficient appliances are given an A+++ rating, while the least efficient appliances are in the D band.

Manufacturers continually change products to deliver improved energy efficiency, while improving effectiveness and reliability. For example, in a washing machine, changes could include:

- high efficiency motors
- greater insulation
- more water-efficient wash cycles
- larger drum sizes.

Using more efficient appliances:

- reduces the electrical energy needed to run the appliance
- reduces the money spent on electricity bills
- reduces the amount of fossil fuels needed to generate the electricity.

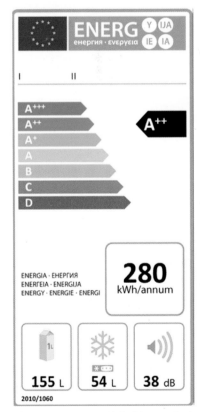

Figure 2.5 *Energy label from a fridge.*

Figure 2.6 *Smart energy meter.*

The UK Government plans for every home and business to have a smart meter for electricity and gas by the end of 2020. Smart meters will keep track of the energy you use in your home and help you use it more efficiently. You can reduce the amount of electricity you use at home by:

- turning lights off when there is no-one in the room

- removing chargers from plug sockets when they are not being used

- turning off all electrical appliances when they are not being used

- turning down the hot water thermostat

- insulating walls and roof space, and reducing draughts around windows and doors to reduce heat loss

- drying your clothes outside instead of using a tumble dryer.

GO! Activity 2.7

☺ 1. Suggest a reason why people do not always buy the most energy-efficient appliances available.
2. Why do you think using energy-efficient appliances helps to reduce global warming?
3. Think of five ways you could reduce the amount of electricity you use.
4. What appliances are left on at your school? Are they left on at night or over the weekend?

National 4

Calculating efficiency

Power or energy efficiency problems can be tackled using these relationships:

$$\text{percentage power efficiency} = \frac{\text{useful power output}}{\text{total power input}} \times 100\%$$

$$\text{percentage energy efficiency} = \frac{\text{useful energy output}}{\text{total energy input}} \times 100\%$$

Table 2.3 *Symbols and units for the efficiency relationships.*

Name	Unit	Unit symbol
Power output and input	watts	W
Energy output and input	joules	J
Percentage efficiency	percent	%

Example 2.5

Calculate the efficiency of a system whose input energy is 30J and transfers 10J to useful energy.

Total energy input = 30J

Useful energy output = 10J

Percentage efficiency = ? — Write out what you know from the question.

$$\text{percentage efficiency} = \frac{\text{useful energy output}}{\text{total energy input}} \times 100\%$$ — Write out the equation.

$$\text{percentage efficiency} = \frac{10}{30} \times 100\%$$ — Substitute in what you know.

percentage efficiency = 33%

Example 2.6

A machine has an input power of 200W. It is 50% efficient. Calculate the output power.

Total power input = 200W

Percentage efficiency = 50%

Useful power output = ? — Write out what you know from the question.

$$50 = \frac{\text{useful power output}}{200} \times 100\%$$

$$\frac{50}{100} = \frac{\text{useful power output}}{200}$$

$$\frac{50}{100} \times 200 = \text{useful power output}$$

useful power output = 100W — Don't forget to write the units in your answer.

GO! Activity 2.8

1. A motor uses 1000J of energy but only gives out 800J to lift a box. Calculate the efficiency of the motor.
2. Calculate the efficiency of an appliance with input energy of 30000J and 1500J of useful output energy.
3. A machine that is 60% efficient has an input power of 1200W. Calculate the output power.
4. The output power from a machine is 2400W. The machine is 30% efficient. Calculate the input power.
5. The power input of a small wind turbine was 24W. The turbine produces an output power of 8W. Calculate the efficiency of the wind turbine.

(continued)

6. A power station is supplied with 100 MJ of energy from the fuel. The output energy is 55 MJ. Calculate the efficiency of the power station.

7. There are different types of light bulbs that transfer different amounts of energy to light.

An incandescent light bulb takes 60 J of energy in and gives 3 J of light energy out.

(a) Calculate the efficiency of the incandescent light bulb.

An energy-saving light bulb takes in 60 J of energy and gives out 50 J of light energy.

(b) Calculate the efficiency of the energy-saving light bulb.

(c) What is the difference between the efficiencies?

(d) What is the wasted energy from a light bulb?

Make the link

A prefix goes before the unit of measurement to help write very large or very small numbers. There is information on prefixes and significant figures in the Investigation Skills Appendix.

Table 2.4 *Table of prefixes.*

Prefix	Symbol	Multiply or divide by	Place value
mega-	M	× 1 000 000	million
kilo-	k	× 1000	thousand
milli-	m	÷ 1000	thousandth
micro-	μ	÷ 1 000 000	millionth

Learning checklist

After reading this chapter and completing the activities, I can:

N3 L3 N4 L4

- state that there is difference in power consumption between common electrical appliances. **Activity 2.1** ○ ○ ○

- state that power is a measure of the energy transferred every second. **Activity 2.1 Q3, Activity 2.2, Activity 2.3** ○ ○ ○

- use the relationship between power, energy and time. **Activity 2.4** ○ ○ ○

- state that the cost of electrical energy can be calculated using utility bills. **Activity 2.5, Activity 2.6** ○ ○ ○

- describe how being energy-efficient can conserve resources and the environment. **Activity 2.7** ○ ○ ○

- use the relationship between input and output to calculate efficiency for power and energy. **Activity 2.8** ○ ○ ○

3 Electromagnetism

This chapter includes coverage of:

N3 Electricity • N3 Forces • N4 Electromagnetism
• Forces, electricity and waves SCN 4-08a

You should already know:

- that magnets can attract one another and attract some metals
- that magnets have a north pole and a south pole
- that like poles repel, unlike poles attract
- that the magnetised needle in a compass points north.

Magnets and magnetic fields

Learning intentions

In this section you will:
- learn how to detect magnetic field patterns with iron filings
- learn that like magnetic poles repel
- learn that unlike magnetic poles attract
- describe the magnetic field patterns for permanent magnets
- investigate the strength of a bar magnet
- state what supermagnets are
- investigate the uses of supermagnets.

| National 3 |
| National 4 |
| Curriculum level 4 |
| Forces, electricity and waves: Forces SCN 4-08a |

When a magnet is placed near an iron nail, the nail experiences a pulling force and moves toward the magnet.

The pulling force gets stronger as the magnet is moved closer to the nail, and gets weaker as the magnet is moved further away from the nail. At a certain distance away, it will not move toward the magnet at all. The strength of the force depends on the **distance** between the magnet and the nail.

The region around a magnet where the nail experiences a force is called a **magnetic field**.

Figure 3.1 *Iron nails experience an attractive force toward a magnet.*

📖 Word bank

- **Magnetic field**

The magnetic field is the region of space around a magnet where other magnetic materials experience a force.

📖 Word bank

- **Permanent magnet**

A permanent magnet is made from a magnetised material and creates its own permanent magnetic field.

Figure 3.3 *Many hill walkers use a compass like this to guide them through the mountains.*

❓ Did you know ...?

The Cuillin Ridge in Skye provides one of the most demanding mountaineering experiences in Scotland. The rock on the ridge is magnetic, so navigating by compass is difficult.

Figure 3.4 *The magnetic gabbro rock in the Cuillin Ridge affects compass navigation.*

❓ Did you know ...?

Magnetic materials called lodestones were first discovered in ancient Greece around 2500 years ago. The word *magnet* in Greek meant 'stone of Magnesia', after the region of ancient Greece where the stones were discovered.

Figure 3.2 *Chinese sailors used lodestones as compasses in ships in the twelfth century.*

Permanent magnets

Permanent magnets keep their magnetism for a long time. In classroom experiments, you will use **bar magnets** and **horseshoe magnets**. These are both examples of permanent magnets. Lodestone is a permanent magnet and it can remain magnetic for thousands of years.

Permanent magnets are commonly made by magnetising metals such as iron, cobalt or nickel. Steel is mostly made of iron, so it can also be magnetised. These metals will also be attracted to other magnets. Many common metals such as copper and aluminium are not magnetic. Precious metals such as silver, gold and platinum are also not magnetic.

A compass contains a permanent magnet that points toward north. The end of the compass that points to north is called a north-seeking pole. The other end of the magnet is called the south-seeking pole, and points south. The names are usually shortened to **north pole** and **south pole**.

Magnetic field lines

The magnetic field around a bar magnet can be seen by scattering iron filings around the magnet.

The iron filings shown in Figure 3.5 align up in lines, known as **magnetic field lines**.

The arrow on a plotting compass, also shown in Figure 3.5, points from North to South. The compass arrow shows the direction of the magnetic field lines and point from north pole to south pole.

The strongest region of a magnetic field is at the poles. At these points, the field lines are closest together and the magnetic force is strongest. The magnetic field gets weaker the further you get from the magnet, and the magnetic force gets weaker. This is shown in Figure 3.6, where the field lines are further apart at a greater distance from the magnet.

When the end of a bar magnet moves close to the end of another bar magnet, it will either be **attracted** or **repelled** (pushed apart).

A north pole and a south pole will attract each other, but two north poles (or two south poles) will repel. When poles attract, the magnetic field lines meet in the middle and appear to flow from one pole to the other. When poles repel, the magnetic field lines never meet and appear to push away from each other.

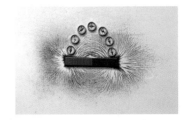

Figure 3.5 *The shape of a magnetic field pattern can be seen using iron filings or a plotting compass.*

Unlike poles attract, like poles repel.

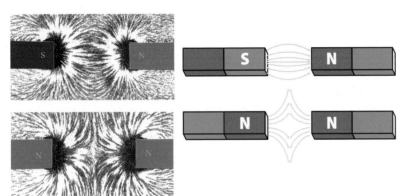

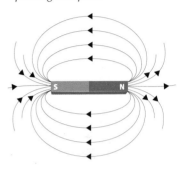

Figure 3.6 *Magnetic field lines drawn around a bar magnet.*

Figure 3.7 *Magnetic field lines between like and unlike poles.*

A horseshoe magnet is a u-shaped bar magnet. The poles are next to one another, so the strength of the magnet is increased. The field lines are much closer together between the poles of a horseshoe magnet, so it can pick up heavier objects.

Figure 3.8 *The magnetic field lines between the poles of a horseshoe magnet.*

Magnets attract and repel other magnets, but they also attract magnetic materials such as iron nails, steel paper clips and drawing pins. Magnetic materials are never repelled and are attracted to either pole or both poles.

The easiest way to make a magnet is to stroke a piece of iron or steel, like a nail, with a magnet. When it is stroked in the same direction, with the same pole, the domains in the nail align themselves in one direction.

Figure 3.9 *Stroking a nail with the same magnetic pole will make it magnetic.*

GO! Activity 3.1

 1. Experiment In this activity you will use a selection of different types of magnets so you can compare different ways of measuring the strength of a magnet.

Read the plan carefully then follow the instructions, making sure you follow all of the safety rules as you would normally be expected to do when carrying out practical work.

You can work in small groups but you should record your own observations and complete the answers yourself.

You will need: a range of permanent magnets, some drawing pins, a shaker of iron filings and a sheet of paper.

Method

(a) For each magnet that you test, use these three methods for determining the strength:

(i) Count the number of drawing pins the magnet can hold. Add pins one at a time to the bottom of the magnet. A stronger magnet will be able to hold more pins. Record the result for each magnet in a table.

Count the number of pins the magnet can hold on to.

(ii) Place a paper clip beside the 0 cm mark on a ruler. Gradually move the magnet along the ruler closer to the paper clip until the paper clip moves toward the magnet. A strong magnet will attract the paper clip at a greater distance than a weak magnet. Use the table below to record the distance the magnet is when the paper clip moves.

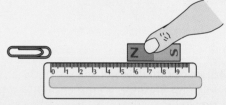

Measure the distance between the paper clip and the magnet with a ruler.

	Number of paper clips held	Minimum distance before paper clip moves toward magnet (cm)
Magnet 1		
Magnet 2		
Magnet 3		

(iii) Place a piece of paper over the magnet, and scatter iron filings over the paper to see the magnetic field lines. The closer the iron filings are together, the stronger the field. Take a photograph of the magnetic field represented by the iron filings for each magnet.

> ⚠ **Beware!**
>
> Be careful not to get iron filings on the magnet itself. They can be very difficult to remove!

(b) From the results, is it possible to say which magnet is the strongest?

(continued)

(c) Which of the methods above was the most conclusive in determining which magnet was the strongest?

(d) What improvements could be made to the experiment to improve the reliability of the results?

2. An iron nail can be made magnetic by stroking it with a permanent magnet. Design an experiment to determine how the number of times it was stroked with a permanent magnet affected the magnetic strength of the nail.

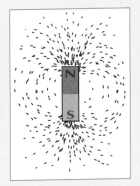

Iron filings moving on a piece of paper placed over a magnet.

⚙ Make the link

Improving reliability in an experiment is explained in the Investigation Skills Appendix.

❓ Did you know ...?

Magnetic north does not align with the geographic North Pole. Magnetic north has been moving slowly to the northwest for approximately 200 years.

The core of the Earth is made mostly of iron, surrounded by a molten outer core. The magnetic field of the Earth is believed to be created by electric currents in the molten core. The Earth's magnetic field is responsible for the Aurora Borealis, often seen in northern parts of Scotland in winter. Charged particles from the Sun enter the atmosphere and are attracted to the North Pole. As the particles accelerate toward the pole they collide with gas molecules in the atmosphere and give off a fantastic show of natural light.

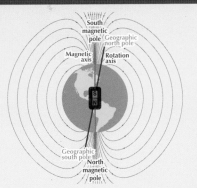

Figure 3.10 *The magnetic field of the Earth does not align exactly with the geographic North and South Pole.*

⚙ Make the link

Find out more about the structure of the Earth in Chapter 17.

Figure 3.11 *The Aurora Borealis over Portobello beach in Edinburgh.*

Figure 3.12 *Neodymium spheres arranged in a cube.*

📖 Word bank

• **Supermagnet**

Supermagnets are the strongest type of permanent magnets currently made. They are made from neodymium, iron and boron.

Supermagnets

In the early 1980s a new type of permanent magnet called neodymium rare earth magnets were created. Known as **supermagnets**, these are the strongest types of permanent magnet in the world.

GO! Activity 3.2

☻ Use these questions to review your understanding of magnetism.

1. What is a magnetic field?
2. Name three materials that can produce a magnetic field.
3. What are the two poles of a permanent magnet called?
4. Two bar magnets are positioned beside one another. They move apart. Describe how the poles of the magnets are arranged.
5. Two bar magnets are positioned beside one another. They move together. Describe how the poles of the magnets are arranged.
6. Draw the magnetic field pattern around:

   (a) a bar magnet

   (b) a bar magnet with its north pole facing the south pole of another bar magnet.
7. How do the magnetic field lines around a permanent magnet show that the magnet is strongest at the poles?

GO! Activity 3.3

☻ Search online to find out what makes supermagnets so powerful. Supermagnets are used in the following applications:

• MRI machines
• computer hard drives
• electric motors
• audio speakers
• electric guitars
• race car engines.

Choose two applications from the list above, and try to find out what function the supermagnet has in their operation.

Electromagnetism

Learning intentions

In this section you will:

- learn what an electromagnet is
- learn what a solenoid is
- describe and draw the magnetic field patterns for a linear solenoid
- investigate what affects the strength of an electromagnet
- describe how the strength of an electromagnet can be investigated.

When a coil of wire is connected to a battery, the current that flows in the wire creates a magnetic field around the coil. A coil used in this way is known as an **electromagnet**.

The coil in Figure 3.13 is called a **linear solenoid**.

The magnetic field created in a linear solenoid is a similar shape to a permanent magnet. There is a north pole at one end of the coil, and a south pole at the other. There are two key advantages of an electromagnet compared to a permanent magnet:

- The wire does not have to be a magnetic material. It is usually made of copper, which is not magnetic, but is a good conductor of electricity.

- The electromagnet stops being magnetic when the current is switched off.

An electromagnet can be made stronger by adding an iron core to the coil of wire. The iron core concentrates the magnetic field around it, making it stronger. An electromagnet with an iron core can produce a magnetic field thousands of times stronger than one without a core. An electromagnet without an iron core is not very effective.

The strength of the field can also be varied by:

- adjusting the number of turns of wire around the iron core

- changing the current flowing through the wire.

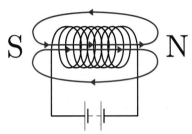

Figure 3.13   *A simple electromagnet. The current flowing through the coil of wire creates a magnetic field.*

📖 Word bank

- **Linear solenoid**

A linear solenoid is a coil of wire that generates a magnetic field when a current flows through the coil.

❓ Did you know ...?

Electromagnets are commonly used for lifting heavy iron objects. Since their magnetism can be turned off instantly, they can carry a load and quickly release it.

This makes them extremely useful in the construction and manufacturing industries. They are also commonly used in scrap yards.

Figure 3.14   *This industrial electromagnet can lift up to 400 kg of iron.*

Activity 3.4

1. Experiment A simple electromagnet can be made by wrapping a thin wire around a large iron nail.
The magnetic field can be made stronger by:

- increasing the current in the wire
- increasing the number of turns of wire around the nail.

In this activity you will plan and carry out an experiment to investigate the effect on the strength of the magnetic field of an electromagnet by changing the number of turns of wire around a nail. You will demonstrate the strength of the magnetic field by counting the number of metal objects (such as pins or paper clips) the nail can pick up.

Read the plan carefully then follow the instructions, making sure you follow all of the safety rules as you would normally be expected to do when carrying out practical work.

You can work in a small group but you should record your own observations and complete the answers yourself.

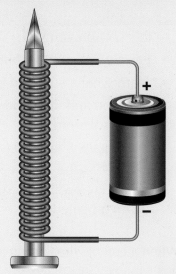

An electromagnetic iron nail picking up pins.

Method

(a) Make up a plan outlining how you will carry out an experiment to investigate how the strength of the electromagnet is affected by the number of coils of wire around an iron nail.

Your plan must include a list of the equipment you will need, and a labelled diagram of the apparatus. State what measurements you will need to take and how you will record and present your results.

(b) State what variables must be kept constant in this experiment.

(c) Ask your teacher to check your plan before you carry out the experiment.

⚠ The wire can become hot when conducting this experiment. Do not leave the wire connected to the power supply for long periods of time.

(d) Record your results in a table. Use the results to draw a graph of the number of metal objects (e.g. pins) lifted against the number of turns of wire.

(e) State any conclusions you can make from the graph.

(f) Discuss with your group any improvements you could have made to the procedure to improve the experiment to get more reliable results.

(g) With your group, come up with a plan outlining how you would investigate how the electric current flowing in the wire affects the strength of the electromagnet. What measures would you have to take to ensure this experiment was carried out safely?

N3 | L3 | **N4** | L4

Generating electricity

Learning intentions

In this section you will:

- investigate how electricity is generated using magnets
- learn what affects the voltage generated by an electric generator
- identify the energy transfer in an electric generator.

Electricity and magnetism are very closely connected. A magnetic field is produced around a wire carrying an electric current. However, if a permanent magnet is moved through a coil of wire, a voltage will be generated across the ends of the wire.

Electrical energy cannot be created without work being done. The magnet must be moving for a voltage to be generated. The size of the voltage generated depends on:

- the **strength of the magnetic field** – a stronger permanent magnet will generate a higher voltage

- the **speed of movement** of the magnet – the faster the magnet moves, the higher the voltage generated

- the **number of turns of wire** on the coil – the greater the number of turns of the coil, the higher the voltage generated.

A voltage will always be generated across a wire if the wire experiences a **changing magnetic field**. This means if the wire moves around a stationary magnet, a voltage will also be generated.

> **Make the link**
>
> Generating electricity is explored in Chapter 1, and more about voltage can be found in Chapter 4.

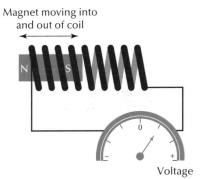

Magnet moving into and out of coil

Voltage

Figure 3.15 *Moving a magnet in and out of a coil of wire generates electricity.*

GO! Activity 3.5

😊😊 **1. Experiment** In this activity you will investigate the voltage generated across a coil of wire using a strong permanent magnet and a school solenoid.

Read the plan carefully then follow the instructions, making sure you follow all of the safety rules as you would normally be expected to do when carrying out practical work.

You can work in a small group but you should record your own observations and complete the answers yourself.

Follow the guidelines detailed below in the method and record your findings.

(continued)

You will need: a solenoid with 125/500 turns of wire, a permanent magnet, a stronger permanent magnet (for example, a neodymium magnet), an analogue voltmeter.

Method

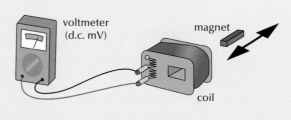

(a) Connect the apparatus as shown.

(b) Hold the magnet stationary inside the coil of wire with 500 turns. What is the reading on the voltmeter? Can you suggest a reason for this?

(c) Move the magnet slowly in and out of the coil of wire, observing the needle on the voltmeter. Now move the magnet quickly in and out of the coil. What is the effect on the voltage produced when you change the speed of the magnet?

(d) Replace the coil with one with fewer turns. How does the voltage produced compare with the previous coil when the magnet is moved in and out of the coil quickly? Suggest a reason for this.

(e) Finally, try moving a stronger magnet in and out of the coil quickly. Does this increase or decrease the voltage produced? Suggest a reason for the change in voltage.

In an electric generator, electricity is produced either

- when a magnet is rotated inside a coil of wire, or

- when a coil is rotated inside a magnet.

The electric generator produces electricity by converting the kinetic energy of the rotating magnet (or coil) into electrical energy. The source of the kinetic energy depends on the type of generator and what it is used for.

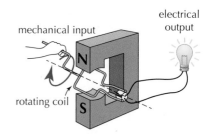

Figure 3.16   *A rotating coil inside a magnet converts kinetic energy into electrical energy.*

☀ Physics in action: Generators large and small

A bicycle dynamo is used to provide electricity for the lamps attached to the bike. The top of the dynamo rubs against bicycle wheel as it turns. A permanent magnet in the dynamo rotates inside the coil of wire. This creates a moving magnetic field that generates a voltage in the coil. Charge flows through the wires to the lamp, to provide lighting.

On a much larger scale, electricity is generated in power stations in a similar way. Hydro, wind, nuclear, oil and gas powers stations all use generators that work on the same principle.

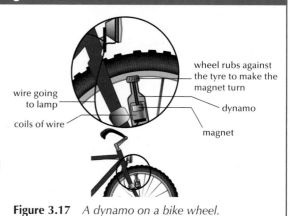

Figure 3.17   *A dynamo on a bike wheel.*

(continued)

There are some important differences between a bike dynamo and a power station generator:

- In a bike dynamo, electricity is generated with a permanent magnet rotating within a coil of wire.

- In a power station, electricity is generated with an electromagnet rotating within a coil of wire. The electromagnet produces a stronger magnetic field than a permanent magnet and is more compact than a permanent magnet would be.

- The electromagnet is driven by a turbine, which is rotated by the kinetic energy of high-pressure steam, water or wind, depending on the type of power station. The kinetic energy of movement generates large amounts of electrical current.

Make the link

See Chapter 1 for more about the generation of electricity in power stations.

Figure 3.18 *Two generators in a power station. They have thousands of turns of wire. They can produce currents of around 20 000 amps.*

GO! Activity 3.6

☻ Use these questions to test your understanding of electromagnetism.

1. State what is meant by an electromagnet.

2. A crane operator in a scrap yard tries to pick up a damaged car using an electromagnet, but discovers the electromagnet is not strong enough.

   (a) What could he do to make the electromagnet stronger?

   (b) What must he do to release the car?

3. An electromagnet is designed as shown.

   (a) What is the name of this type of electromagnet?

   (b) Copy and complete the diagram to show the magnetic field pattern around this electromagnet.

   (c) Suggest three changes that could be made to increase the strength of the electromagnet.

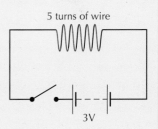

5 turns of wire

3V

4. A magnet is placed next to a coil of wire.

   (a) State what must be done to the magnet to generate an electric current in the coil.

   (b) State three ways in which the amount of current could be increased in the coil.

   (c) State the energy conversion in an electric generator.

5. A cyclist finds that the lamp connected to the dynamo on their bicycle does not glow very brightly when cycling uphill. However, it glows very brightly going downhill.

   (a) State and explain the reason for this.

   (b) Suggest two improvements the manufacturer of a dynamo could make to increase the voltage generated from the dynamo.

Practical applications of magnets and electromagnets

Learning intention

In this section you will:

- learn about applications of permanent magnets and electromagnets.

Permanent magnets and electromagnets have a variety of applications. In many cases, permanent magnets are suitable, but there are situations where it makes more sense to use an electromagnet.

Electric motors

A simple electric motor turns electrical energy into kinetic energy. An electric motor contains a coil of wire that can rotate between two opposite poles of a permanent magnet. When an electric current passes through the coil, it experiences a force in the magnetic field and rotates.

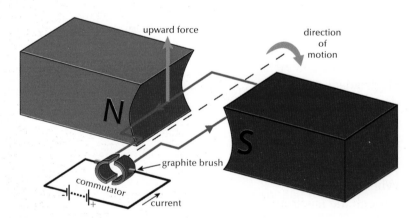

Figure 3.19 *The direction of the current is reversed every half turn by the commutator to allow the wire to keep spinning.*

Electric motors are used in many different applications. In homes, electric motors are found in vacuum cleaners, washing machines and tumble dryers, as well as electronic equipment like DVD players and children's toys.

GO! Activity 3.7

☺ **1.** Copy the table below. Create as many rows as there are rooms in your house.

Room in your home	Appliance with an electric motor

Starting in the kitchen, write down every appliance you can find in each room that has an electric motor.

If you have an outbuilding like a garage or shed then look in there too. Many power tools have electric motors.

> 🔍 **Hint**
>
> If an appliance has an electric motor, it will have a rotating part.

2. How many appliances in your home have electric motors?

3. Cars have a number of electric motors. Make a list of parts of a car that use an electric motor.

4. Search online to research applications of permanent magnets in the home. Make a list of applications of permanent magnets, stating which room in the home they are likely to be found. How many on the list can you find in your home?

The electric bell

Electromagnets are used in many electrical devices, and are sometimes used alongside permanent magnets.

Figure 3.20 shows a circuit diagram for an electric bell.

When the switch is closed, current flows through the electromagnet. This magnetises it, pulling the hammer up to hit the bell. This also causes the contacts to pull apart, breaking the circuit. This turns the electromagnet off, causing the hammer to spring away from the bell. The contacts then reconnect, allowing current to flow again through the electromagnet. The hammer then hits the bell again. This process repeats until the switch is released.

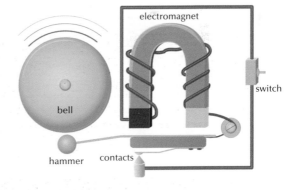

Figure 3.20 *An electric bell.*

The relay

Relays are a type of switch. They are used when a small current is needed to control a circuit with a much larger current, such as in a washing machine motor at home.

> ⚗️ **Make the link**
>
> The relay output device is explained in Chapter 7.

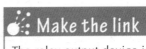

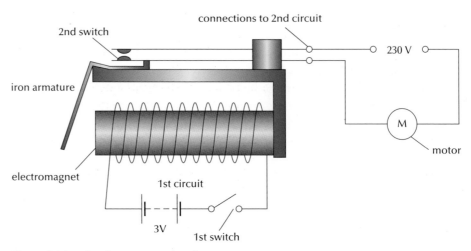

Figure 3.21 *An electromagnetic relay.*

When the first switch is closed, a small current flows through the electromagnetic. This attracts the iron armature toward it, pushing closed the second switch. The second circuit is completed and the motor runs.

The loudspeaker

Loudspeakers use both electromagnets and permanent magnets.

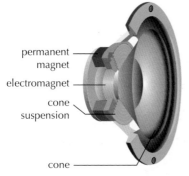

Figure 3.22 *The internals of a loudspeaker.*

The electromagnet sits between the poles of a permanent magnet. The incoming electrical signal to the loudspeaker varies, causing the electromagnet to magnetise one way and then another. This pushes it in and out of the permanent magnet, moving the cone along with it, at the same frequency as the music.

⁂ Physics in action: Large-scale electromagnets

Electromagnets are very powerful and have many important industrial, medical and scientific uses.

Levitating trains

Maglev trains use magnetic levitation to move without touching the tracks. The train can travel along the tracks using electromagnets to create lift, reducing the friction on the track. This allows them to travel much faster than conventional trains.

The Shanghai Maglev train is the fastest in the world, reaching a top speed of 270 mph.

Figure 3.23 *The Shanghai Maglev train.*

Seeing inside the body

An advanced method for imaging inside the human body is called Magnetic Resonance Imaging (MRI). MRI scanners use strong magnetic fields produced by powerful electromagnets. These magnetic fields can map the location of water and fat in the body. Water accounts for about 57% of the total mass of the human body, and by mapping the location and distribution of water molecules, it is possible to create very detailed images of organs, bones, muscles and other tissue within the human body.

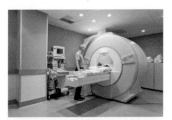

Figure 3.24   *An MRI scanner.*

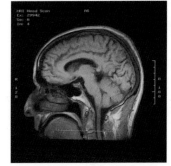

Figure 3.25   *An MRI scan of a patient's head.*

Seeing inside the atom!

The Large Hadron Collider (LHC) is the world's largest and most powerful particle accelerator. It is part of the European Organization for Nuclear Research, based at the CERN laboratory in Geneva.

The LHC accelerates particles close to the speed of light, and then allows them to collide. The particles are guided around the 27 km ring by a strong magnetic field created by powerful electromagnets. The LHC has made amazing discoveries about the fundamental particles that make up the universe. In 2013, scientists at the LHC confirmed the existence of the Higgs boson. Professor Peter Higgs, of Edinburgh University, first proposed the existence of the Higgs boson in 1964, but it took nearly 50 years for the experimental proof to be demonstrated.

Figure 3.26   *A section of the 27 km loop of the particle accelerator.*

GO! Activity 3.8

😊😊 Create a poster to describe applications of magnets and electromagnets that includes the following:

- three applications that use permanent magnets
- three applications that use electromagnets
- one application that has had an important industrial, medical or scientific use.

Make the link

Using posters as a form of scientific communication is explained in the Investigation Skills Appendix.

National 4

Using transformers in high voltage transmission

📖 Word bank

- **Alternating current (a.c.)**

When an electric current is made to flow in one direction and then another repeatedly, it is known as alternating current (a.c.). All mains electrical supplies use alternating current.

A transformer is a device that can change the value of an **a.c.** supply.

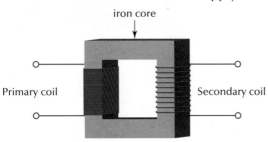

iron core

Primary coil

Secondary coil

Figure 3.27   *A transformer.*

A transformer has three main parts:

- the **primary coil**, connected to an input a.c. supply
- the **secondary** coil, not connected to a voltage supply
- an **iron core**.

The a.c. supply connected to the primary coil causes an **alternating current to flow in the primary coil**.

↓

The alternating current in the primary coil creates a **changing magnetic field in the primary coil**.

↓

The iron core transmits the **changing magnetic field to the secondary coil**.

↓

The changing magnetic field in the secondary coil creates an **alternating voltage in the secondary coil**.

↓

This voltage is the **output** of the transformer.

The **voltage output** can be controlled by adjusting the **number of turns of wire** in the primary and secondary coils. A direct current (d.c.) supply cannot be used with a transformer as it does not create a changing magnetic field.

The voltage output of a transformer depends on the number of turns of wire in both the primary and secondary coils.

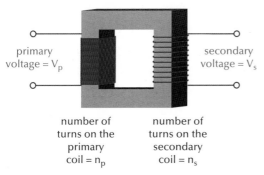

primary
voltage = V_p

secondary
voltage = V_s

number of
turns on the
primary
coil = n_p

number of
turns on the
secondary
coil = n_s

Figure 3.29 *Primary and secondary coils on a transformer.*

The relationship between the turns of wire in each coil and the primary and secondary voltages is given by the equation shown here.

$$\frac{n_s}{n_p} = \frac{V_s}{V_p}$$

where:
n_p = number of turns of wire in the primary coil
n_s = number of turns of wire in the secondary coil
V_p = voltage across the primary coil
V_s = voltage across the secondary coil.

> ★ **You need to know**
>
> Transformers can either make the input voltage bigger or smaller:
>
> - a **step-up transformer** makes the output voltage **bigger** than the input voltage
>
> - a **step-down transformer** makes the output voltage **smaller** than the input voltage.

> ? **Did you know ...?**
>
> Chargers for mobile phones contain transformers which reduce the high voltage mains supply to a voltage appropriate for charging the phone.
>
>
>
> **Figure 3.29** *A smartphone charger uses a transformer.*
>
> The charger steps down the voltage from 230V to around 5V.

Example 3.1

The primary voltage applied to the transformer in a mobile phone charger is 230V. The number of turns of wire in the primary coil is 5000. The voltage output from the transformer is 5V (this is the secondary voltage). Calculate the number of turns in the secondary coil.

$V_p = 230V$

$n_p = 5000$ ———○————(Write out what you know from the question.)

$V_s = 5V$

$n_s = ?$

$\dfrac{n_s}{n_p} = \dfrac{V_s}{V_p}$ ———○————(Write out the equation.)

(continued)

$$\frac{n_s}{5000} = \frac{5}{230}$$ — Substitute in what you know and solve for n_s.

$$n_s = \frac{5}{230} \times 5000$$

$n_s = 109$ turns of wire.

Example 3.2

A microwave oven uses a step-up transformer to generate the high voltage required to produce microwaves. The transformer has 50 turns of wire in the primary coil and 1000 turns of wire in the secondary coil. A primary voltage of 230V is applied to the transformer. Calculate the voltage output from the secondary coil.

$V_p = 230V$

$n_p = 50$

$V_s = ?$

$n_s = 1000$ — Write out what you know from the question.

$$\frac{n_s}{n_p} = \frac{V_s}{V_p}$$ — Write out the equation.

$$\frac{1000}{50} = \frac{V_s}{230}$$ — Substitute in what you know and solve for V_s.

$$V_s = \frac{1000}{50} \times 230$$

$V_s = 4600\,V$ — Don't forget to write the units in your answer.

 Make the link

Transformers are used to step up the voltage output from a power station to save energy when transmitting it around the country. This is explained in Chapter 1.

GO! Activity 3.9

1. Experiment In this activity you will investigate the relationship between the primary and secondary voltages and the primary and secondary coils.

Read the plan carefully then follow the instructions, making sure you follow all of the safety rules as you would normally be expected to do when carrying out practical work.

You can work in a small group, but you should record your own observations and complete the answers yourself.

You will need: two voltmeters, a variable a.c. voltage power supply, two transformer coils with a range of different numbers of turns and an iron core. Common coil turns found in school laboratories are 60/60 or 125/500 turn coils.

Method

(a) Connect your apparatus as shown.

(b) Copy the table below. The values for the primary and secondary turns will depend on what coils there are available to you. The first row has been completed for you.

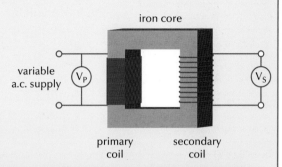

Turns in primary coil (n_P)	Turns in secondary coil (n_S)	$\dfrac{n_s}{n_P}$	Primary voltage (V_P) (V)	Secondary voltage (V_S) (V)	$\dfrac{V_S}{V_P}$
60	120		2.5		
			3		
			1		
			6		
			4		
			2		

(c) To complete the first row of data in the table, collect two 60/60 turn coils. Make the primary coil (n_p) be 60 turns and the secondary coil (n_s) be 120 turns. Connect the primary coil to the variable a.c. supply and connect a voltmeter across the primary coil also. Set the voltage across the primary coil (V_p) to 2.5V. Connect a voltmeter across the secondary coil and measure the voltage. Record this voltage (V_s) in the first row of the table.

(d) Repeat step (c) for the five other rows in the table by altering the primary and secondary coils and setting the primary voltage to the value shown in the table.

(e) Calculate the ratios $\dfrac{n_s}{n_p}$ and $\dfrac{V_s}{V_p}$ and record the results in the table.

(f) What do you notice about the values for $\dfrac{n_s}{n_p}$ and $\dfrac{V_s}{V_p}$ in the table? How do these columns confirm the relationship for an ideal transformer?

GO! Activity 3.10

☻ **1.** Copy and complete the following paragraph using the words provided to describe transformers and their uses.

step-up   **mains**   **decreases**   **changing**
alternating   **iron core**   **secondary**   **input**

A transformer consists of two coils of wire wrapped around an _ _ _ _ _ _ _ _.
The _ _ _ _ _ coil is called the primary coil. The output coil is called the _ _ _ _ _ _ _ _ _ _
coil. Transformers work with _ _ _ _ _ _ _ _ _ _ _ current, such as _ _ _ _ _ supply.
There must be a _ _ _ _ _ _ _ _ magnetic field around the coils. A transformer that
increases the input voltage is called a _ _ _ _ _ _ transformer. A transformer
that _ _ _ _ _ _ _ _ _ the input voltage is called a step-down transformer.

2. Use the relationship for an ideal transformer

$$\frac{n_S}{n_P} = \frac{V_S}{V_P}$$

to find the missing values in the following table.

n_S	n_P	V_S (V)	V_P (V)
60	120	6	
125		4	16
5000	500		2.5
	2000	2	10
500	8000	25000	
1000		240	6

3. A welding machine uses a transformer to step down the voltage. The 230V mains supply is connected to 5000 turns in the primary coil. There are 60 turns in the secondary coil. What is the output voltage of the transformer?

4. A mobile phone charger uses a transformer to step down mains voltage from 230V to 5V. There are 50 turns of wire in the secondary coil. How many turns of wire are in the primary coil?

5. The voltage produced by a hydroelectric power station is stepped up by a transformer to 400 000V to be transmitted along overhead cables. The transformer has 2000 turns of wire in the primary coil and 32 000 turns of wire in the secondary coil. What is the voltage produced by the power station?

Learning checklist

After reading this chapter and completing the activities, I can:

N3 **L3** **N4** **L4**

- identify different methods for measuring the strength of a permanent magnet. **Activity 3.1** ○ ○ ○
- state that iron filings can be used to visualise the magnetic field patterns around magnets. **Activity 3.1** ○ ○ ○
- state that supermagnets are the strongest types of permanent magnet in the world. **Activity 3.2** ○ ○ ○
- state applications of supermagnets. **Activity 3.2** ○ ○ ○
- state that like magnetic poles repel. **Activity 3.3 Q4** ○ ○ ○
- state that unlike magnetic poles attract. **Activity 3.3 Q5** ○ ○ ○
- describe the magnetic field patterns between the poles of a magnet. **Activity 3.3 Q6, Q7** ○ ○ ○
- state that an electromagnet is created when a charge flows through a coil of wire. **Activity 3.4** ○ ○ ○
- state that the strength of an electromagnet can be increased by increasing the number of coils of wire, increasing the current in the coils of wire, or by adding an iron core to the electromagnet. **Activity 3.4, Activity 3.6 Q2(a)** ○ ○ ○
- state that electricity is generated by moving a magnet near a coil of wire. **Activity 3.5, Activity 3.6 Q4(a)** ○ ○ ○
- state that the voltage generated by an electric generator depends on the strength of the magnetic field, the speed of movement of the magnetic field, and the number of turns of wire on the coil. **Activity 3.5, Activity 3.6 Q4(b), Q5(b)** ○ ○ ○
- state that a solenoid is a coil of wire that generates a magnetic field when a charge flows through it. **Activity 3.6 Q3(a)** ○ ○ ○
- describe the magnetic field lines around a solenoid. **Activity 3.6 Q3(b)** ○ ○ ○

| N3 | L3 | N4 | L4 |

N3 L3 N4 L4

- state that electric generators convert kinetic energy into electrical energy. **Activity 3.6 Q4(c)**

  ◯ ◯ ◯

- describe different practical applications of magnets and electromagnets. **Activity 3.7, Activity 3.8**

  ◯ ◯ ◯

- state that the output of a transformer depends on the ratio of number of turns of wire in the primary and secondary coils. **Activity 3.9, Activity 3.10 Q2**

  ◯ ◯ ◯

- state that a transformer changes the voltage of an a.c. supply. **Activity 3.10 Q1**

  ◯ ◯ ◯

- use the relationship between the number of coils of a transformer and the primary and secondary voltages to perform calculations based on the number of turns of wire in the coils of a transformer and the input and output voltages of the transformer. **Activity 3.10 Q2–Q5**

  ◯ ◯ ◯

4 Electrical circuits

This chapter includes coverage of:

N3 Electricity • N4 Electricity • Forces, electricity and waves SCN 3-09a • Forces, electricity and waves SCN 4-09a

You should already know:

- how to stay safe when using electricity
- the importance of electricity in our daily lives
- that some objects may become electrically charged by rubbing two surfaces together and that the charges produce an electrostatic force
- that there are two kinds of static charge – positive and negative
- that an electrical circuit is a continuous loop of conducting materials
- how to combine components to make a series circuit
- that electrical components can be represented by symbols
- that energy is transferred around a circuit.

Current in a circuit

Curriculum level 3

Forces, electricity and waves: Electricity SCN 3-09a

Learning intentions

In this section you will:

- explain that current is the movement of electrical charges around a circuit
- learn that current is measured in amps (A)
- learn that current is measured with an ammeter.

An electrical circuit is used to transfer energy from an energy source such as a battery to components such as lamps, heaters and motors.

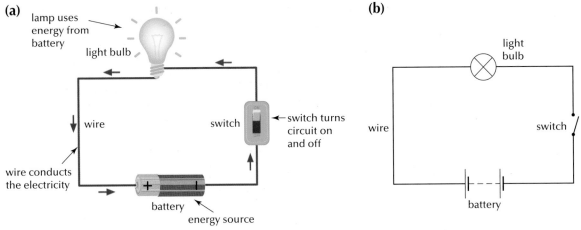

Figure 4.1 *(a) A simple electric circuit (b) Circuit diagram using component symbols.*

The circuit in Figure 4.1 is called a **complete circuit**. When the switch is closed to complete the circuit, there are no breaks in it between the battery and the lamp. When the circuit is complete, the lamp will light. Energy is transferred from the battery to the lamp. The energy is transferred by an **electric current**. The circuit diagram in Figure 4.1(b) uses symbols to represent the light bulb, switch and battery.

📖 **Word bank**

• **Current**

Current is the rate of flow of charge (electrons) in the circuit, and is given the symbol *I*. It is measured by an ammeter in amps (A).

Current

An electric current is simply the flow of electric charges. A metal wire contains **electrons** that are able to move freely about inside the wire. The electrons are negatively charged. When a wire is connected to a battery as part of a complete circuit, the negatively charged electrons are pushed around the circuit in one direction.

📖 **Word bank**

• **Electrons**

Electrons are tiny particles that form part of an atom. They have a negative charge. Current is a flow of electrons in a circuit.

Measuring current

A device called an **ammeter** is used to measure the current in a circuit. As shown in Figure 4.3, the circuit is broken and the ammeter is inserted into the circuit to measure the current.

📖 **Word bank**

• **Ammeter**

An ammeter measures the current flowing in a circuit.

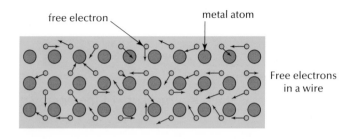

free electron   metal atom

Free electrons
in a wire

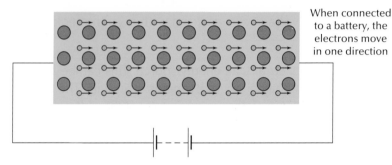

When connected
to a battery, the
electrons move
in one direction

Figure 4.2   *The way the electrons move changes when the wire is connected to a battery.*

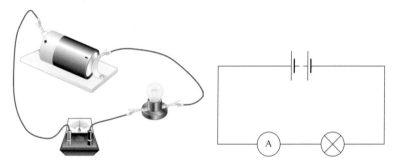

Figure 4.3   *An ammeter connected to measure the current in a circuit.*

? Did you know ...?

The unit of current is the ampere, named after André-Marie Ampère who demonstrated the first ammeter in 1820. The word 'amperes' is usually shortened to 'amps' (A).

Figure 4.4   *Two ammeters that you might find in school, both measuring 0.5 amps.*

GO! Activity 4.1

☺ Copy and complete the following paragraph about current using the words provided:

charges   **current**   **flow**   **components**

ammeter   **amps**   **energy**

_ _ _ _ _ _ _ is a measure of the rate of _ _ _ _ of
_ _ _ _ _ _ _ in a circuit. Current is measured in _ _ _ _.
Current transfers _ _ _ _ _ _ from the battery to the
_ _ _ _ _ _ _ _ _ _ in a circuit. An _ _ _ _ _ _ _ is used to
measure current.

? Did you know ...?

A current of 1 amp means
6 250 000 000 000 000 000
electrons flow past a point every second. Boiling water in a kettle draws about 10 amps. Can you work out how many electrons would flow in and out of your kettle every second?

Figure 4.5   *Boiling a kettle for a cup of tea moves a lot of electrons!*

📖 Word bank

- **Voltage**

  Voltage is a measure of the energy given to electric charges in a circuit, and is given the symbol *V*. The unit of voltage is the volt, which is also abbreviated to the symbol V.

Figure 4.6 *All batteries have a positive and a negative terminal and display their voltage rating.*

Figure 4.7 *A cell and its circuit symbol.*

Figure 4.8 *A battery and its circuit symbol.*

Figure 4.9 *Low voltage power supply.*

Voltage in a circuit

Learning intentions

In this section you will:

- explain that voltage is a measure of the energy given to charges
- learn that voltage is measured in volts (V)
- learn that voltage is measured with a voltmeter.

Electrical circuits require an energy source. This can be provided by a cell, a battery or by mains electricity. The energy that can be provided by the energy source is the **voltage**. The voltage provided by these energy sources has two purposes:

- to 'push' the charges around the circuit
- to supply the energy for the components to operate.

When there is no voltage, no current will flow because nothing is pushing the charges. The larger the voltage of the battery or power supply, the bigger the push and the greater the current.

Every battery has two important pieces of information printed on it. In Figure 4.6, the battery is identified as an alkaline battery, with a voltage of 9 V. The word **alkaline** tells us about the type of chemical reaction that the battery uses to produce electrical energy.

In physics, a single battery (e.g. a 1.5 V alkaline battery) is called a **cell**. A battery is a series of cells connected together.

Electricity experiments at school are carried out with cells, batteries, or **mains voltage** power packs (Figure 4.9). These power packs transform the high voltage of the mains down to a usable low voltage. This makes it safe to use. The voltage output from these power supplies can be controlled so they are useful in experiments where a range of voltages is needed.

In Figure 4.10 a **voltmeter** is connected across a cell. It is measuring the difference in energy between the charges leaving the cell and those returning to it.

Voltage is always measured by connecting a voltmeter across components. Current is measured by breaking the circuit and inserting an ammeter into the circuit, in line with the components.

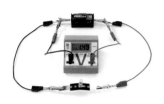 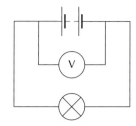

Figure 4.10 *The voltage of a cell is measured by a voltmeter. The circuit diagram shows the voltmeter connected across the cell. The circuit does not have to be broken for a voltmeter to measure the voltage.*

Make the link

Transformers are explained in more detail in Chapter 3.

Word bank

• **Mains voltage**

The voltage supply from the mains electricity coming to a home is 230 V. This voltage supply comes direct from an electric power station.

• **Voltmeter**

A voltmeter measures the energy difference between two points in a circuit.

Activity 4.2

☻ Copy and complete the following paragraph about voltage using the words provided:

charges	voltage	battery	two	voltmeter
mains	energy	cells	volts	

_ _ _ _ _ _ _ is a measure of the _ _ _ _ _ _ given to _ _ _ _ _ _ _ in a circuit. Voltage is measured in _ _ _ _ _. The voltage supply in a circuit is usually provided by a _ _ _ _ _ _ _ or the _ _ _ _ _ supply. A battery is made up of _ _ _ or more _ _ _ _ _ _. A _ _ _ _ _ _ _ _ _ _ is used to measure voltage.

Series and parallel circuits

Learning intentions

In this section you will:

• describe the features of series and parallel circuits and the difference between them

• learn how to measure current and voltage using appropriate meters in series and parallel circuits

• learn that the current in a series circuit is the same at all points in the circuit

• learn that the sum of the voltages across the components in a series circuit is equal to the supply voltage

• learn that the current leaving the supply is equal to the sum of the current through each branch in a parallel circuit

• learn that the voltage across the supply is equal to the voltage across each branch in a parallel circuit.

Curriculum level 3

Forces, electricity and waves: Electricity SCN 3-09a

National 4

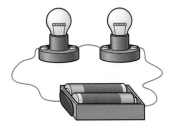

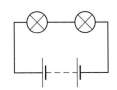

Figure 4.11 *Two lamps connected in series to a battery.*

The way in which components are arranged in a circuit can affect how they operate and how useful they are. The simplest arrangement for components is a series circuit.

Series circuits

In a series circuit,

- all the components are connected one after another, end to end, in one complete loop

- there is only one path that the current can take

- any break in the circuit will cause the whole circuit to stop working as the only path for the current has been broken.

Activity 4.3

1. Experiment In this activity you will investigate the basic operation of a series circuit, and what happens when a fault occurs.

Read the plan carefully then follow the instructions, making sure you follow all of the safety rules as you would normally be expected to do when carrying out practical work.

You can work in a small group but you should record your own observations and complete the answers yourself.

You will need: two 1.5 V lamps, one 1.5 V cell, a switch and four connecting leads

Method

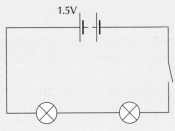

(a) Set up the series circuit as shown, with both lamps connected to the cell.

(b) Close the switch to turn on the lamps. Explain why closing the switch turns the lamps on.

(c) With the switch still closed, unscrew one of the lamps to simulate a fault in the circuit.

(d) Explain why both lamps go out when you unscrew one lamp.

Parallel circuits

Another way to connect components in a circuit is in parallel.

In a parallel circuit,

- components can be connected side by side, separately in their own loop

- there is more than one path for the current to take

- if there is a break in the circuit in one loop, current can still flow through the other loop.

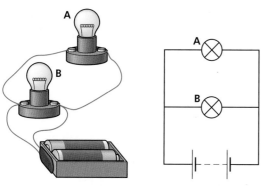

Figure 4.12   *Two lamps connected in parallel to a battery.*

You have already come across a parallel circuit, since a voltmeter is connected in parallel across a component.

Activity 4.4

 1. Experiment In this activity you will investigate the basic operation of a parallel circuit, how switches can be used to control different parts of the circuit and what happens when a lamp breaks.

Read the plan carefully then follow the instructions, making sure you follow all of the safety rules as you would normally be expected to do when carrying out practical work.

You can work in a small group but you should record your own observations and complete the answers yourself.

You will need: one 1.5 V cell, two 1.5 V lamps, three switches and seven connecting leads.

Method

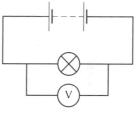

Figure 4.13   *A voltmeter connected **in parallel** across a lamp.*

(a) Set up the circuit as shown with two lamps connected to the cell in parallel.

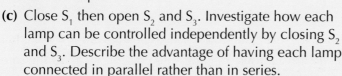

(b) Close all three switches. Open Switch 1 (S_1). What effect does it have on the lamps?

(c) Close S_1 then open S_2 and S_3. Investigate how each lamp can be controlled independently by closing S_2 and S_3. Describe the advantage of having each lamp connected in parallel rather than in series.

(d) With all switches closed, unscrew one of the lamps to simulate a fault in that lamp. Explain why the other lamp stays on.

Traditional Christmas tree lights were connected in series. When one lamp broke, they all went out. This made it incredibly difficult to find the faulty lamp.

Figure 4.14 *Looking for the faulty lamp on the Christmas tree lights.*

Modern Christmas tree lights are connected in parallel. That way, if one lamp breaks, current can still reach the other lamps.

Current and voltage in series and parallel circuits

The circuits in Figure 4.17 show how two identical lamps are connected in series and in parallel. The current in each lamp and the energy they both receive from a voltage supply is very much dependent on whether it is connected in series or in parallel.

(a) **(b)**

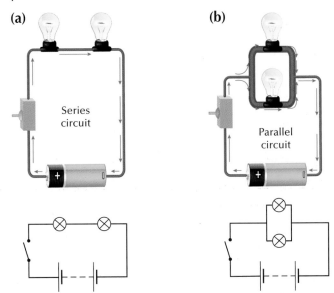

Series circuit

Parallel circuit

Figure 4.15 *Current and voltage are not the same in **(a)** series and **(b)** parallel circuits.*

Rules for current and voltage in a series circuit

Since there is only one route for the current in a series circuit, the current is the same at any part of the circuit. An ammeter will show the same reading wherever it is placed.

$$I_1 = I_2$$

The voltage from the supply is shared between the two lamps. The voltage across each lamp adds up to the total voltage supplied by the cell. Since each lamp is identical, the voltmeter reads 3V across each lamp, dividing the voltage of the cell equally.

$$V_{supply} = V_1 + V_2$$

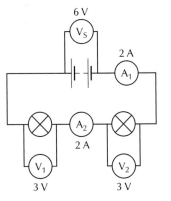

Figure 4.16 *Current and voltage in a series circuit.*

Activity 4.5

1. Experiment In this activity you will investigate how the current varies in a series circuit. Read the plan carefully then follow the instructions, making sure you follow all of the safety rules as you would normally be expected to do when carrying out practical work.

You can work in a small group but you should record your own observations and complete the answers yourself.

You will need: two 6 V lamps, a 6 V battery, a switch, an ammeter and five connecting leads.

Method

    (a) Set up the series circuit shown with two lamps connected to the 6 V battery, a switch and an ammeter to measure the current.

    (b) Close the switch and check that both lamps are lit. Measure the current using the ammeter and record it in a table, next to a description of the position of the ammeter.

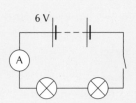

Position of ammeter	Current (A)
Between battery and first lamp	
Between first and second lamp	
Between second lamp and switch	
Between switch and battery	

    (c) Repeat the experiment with the ammeter at different positions in the circuit as shown by the diagrams below.

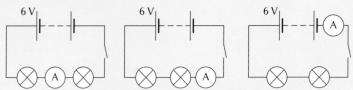

    (d) How does the measured value of current depend on the position of the ammeter?

? Did you know ...?

When a switch is turned on in a circuit, charges flow at every point in the circuit at the same time. This is regardless of how long the circuit is. This is because the electrons in the wire experience the push from the battery almost instantaneously. However, the electrons do not travel around the circuit quickly. For example, the electrons that move in a wire for electric lighting in a home move at around 80 cm per hour, or about 1/5 of a millimetre every second.

Figure 4.17   *A cable used for wiring a house.*

GO! Activity 4.6

1. Experiment In this activity you will investigate how the voltage varies in a series circuit. Read the plan carefully then follow the instructions, making sure you follow all of the safety rules as you would normally be expected to do when carrying out practical work.

You can work in a small group but you should record your own observations and complete the answers yourself.

You will need: a 6 V battery, two resistors of different resistance, a switch, a voltmeter and six connecting leads.

(continued)

N3	**L3**	N4	L4	**71**

Method

(a) Set up the series circuit shown, with the two resistors connected in series to the 6 V battery.

⚠ Remember! Open the switch after each reading to ensure the circuit does not overheat.

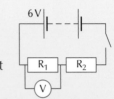

(b) Connect the voltmeter across the first resistor, labelled R_1 in the diagram. Close the switch and measure the voltage across R_1 and record it in a table.

(c) Disconnect the voltmeter from R_1 and use it to measure the voltage across R_2 and the battery. Ensure the switch is closed when the voltage is measured.

Position of voltmeter	Voltage (V)
Across R_1	
Across R_2	
Across battery	

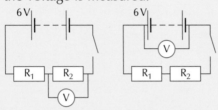

(d) How does the size of the voltage from the battery supply compare with the voltage across each of the resistors?

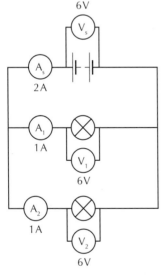

Figure 4.18 *Current and voltage in a parallel circuit.*

Rules for current and voltage in a parallel circuit

Since there is more than one route for current to flow in a parallel circuit, the current splits up between each branch. The current leaving the supply is split in two. Therefore, adding up the current in each branch will give the total current from the cell.

$$I_{supply} = I_1 + I_2$$

The voltage across each branch in a parallel circuit is equal to the voltage provided by the supply. A voltmeter connected across the cell will show the same reading as one connected across each lamp.

$$V_{supply} = V_1 + V_2$$

GO! Activity 4.7

1. Experiment In this activity you will investigate how the current varies in a parallel circuit. Read the plan carefully then follow the instructions, making sure you follow all of the safety rules as you would normally be expected to do when carrying out practical work.

You can work in a small group but you should record your own observations and complete the answers yourself.

You will need: two resistors of different sizes, a 6 V battery, a switch, an ammeter and connecting leads.

Method

(a) Set up the parallel circuit shown with two resistors, the 6 V battery, the switch and an ammeter to measure the current.

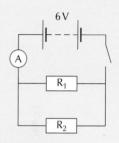

(b) Close the switch and measure the current leaving the battery using the ammeter and record it in a table next to a description of the position of the ammeter.

Position of ammeter	Current (A)
In series with the battery (measuring I_{supply})	
In series with first resistor (measuring I_1)	
In series with second resistor (measuring I_2)	

(c) Repeat the experiment with the ammeter at different positions in the circuit as shown by the diagrams.

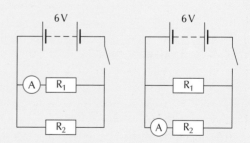

(d) How does the value of the current leaving the battery supply (I_{supply}) compare with the current in each of the two resistors (I_1 and I_2)?

(e) Add up the current in both resistors. How does it compare with the current coming from the battery?

GO! Activity 4.8

 1. Experiment In this activity you will investigate how the voltage varies across components in a parallel circuit.

Read the plan carefully then follow the instructions, making sure you follow all of the safety rules as you would normally be expected to do when carrying out practical work.

You can work in a small group but you should record your own observations and complete the answers yourself.

You will need: two resistors of different sizes, a 6 V battery, a switch, a voltmeter and seven connecting leads.

Method

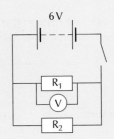

(a) Set up the parallel circuit shown with two resistors connected in parallel. The switch should control the current to both resistors.

(b) Close the switch and measure the voltage using the voltmeter and record it in a table next to a description of the position of the voltmeter.

Position of voltmeter	Voltage (V)
Across the first resistor, V_1	
Across the second resistor, V_2	
Across the battery, V_s	

 Remember! Switch the circuit off between voltmeter readings to ensure the resistors do not overheat.

(c) Repeat the experiment with the voltmeter at different positions in the circuit as shown by the diagrams. Remember to keep the switch closed when measuring the voltage.

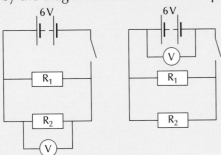

(d) How does the value of the voltage across the battery compare with the voltage across each of the two resistors?

The rules for current and voltage in series and parallel circuits are shown in Table 4.1.

Table 4.1 *Current and voltage in series and parallel circuits.*

Circuit	Current	Voltage
Series	Equal to current leaving supply $I_s = I_1 = I_2 = ...$	Adds up to supply voltage $V_s = V_1 + V_2 + ...$
Parallel	Adds up to supply current $I_s = I_1 + I_2 + ...$	Equal to voltage across supply $V_s = V_1 = V_2 = ...$

Example 4.1

Calculate the missing values for current and voltage for the circuit shown.

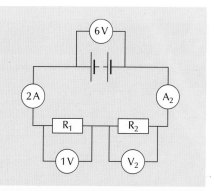

$I_1 = I_2$ • — We know that the current is the same at every point in a series circuit.

$I_2 = 2$ A • — The current through ammeter A$_2$ has been given the symbol I_2.

(continued)

$V_S = V_1 + V_2$ • ━━━ We know that adding up the voltages across each resistor is equal to the voltage from the supply V_s.

$6\,V = 1V + V_2$ • ━━━ Substitute what you know from the question.

$V_2 = 6\,V - 1\,V$ • ━━━ Solve to find the value of V_2.

$V_2 = 5\,V$ • ━━━ Don't forget to include the units.

Example 4.2

Calculate the missing voltage and current values for the circuit shown.

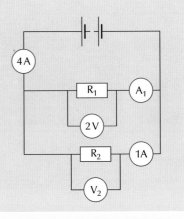

$V_S = V_1 = V_2$ • ━━━ We know that the voltage is the same across every branch in the circuit and equals the voltage of the supply.

$V_2 = 2\,V$ • ━━━ V_2 is the voltage across R_2 and is the same as the voltage across R_1.

$I_S = I_1 + I_2$ • ━━━ The current leaving the battery is split between the branches in a parallel circuit. The current in each branch adds up to the current leaving the supply.

$4\,A = I_1 + 1\,A$ • ━━━ Substitute what you know from the question and solve to find the value of I_1.

$I_1 = 3\,A$ • ━━━ Don't forget to include the units.

GO! Activity 4.9

☻ **1.** The figure shows two circuits with different numbers of cells connected to a lamp.

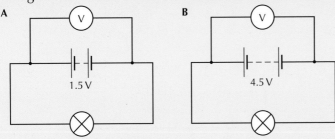

(a) Which lamp will be the brightest? Can you explain your answer?

(b) If another circuit was constructed and only two of the cells were used, what would be the voltage reading on the voltmeter?

2. A student sets up a circuit to allow a lamp to be turned on and off.

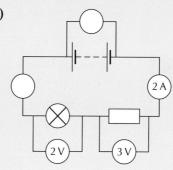

   (a) Redraw the diagram to show where the student should add:

      (i) a voltmeter to measure the voltage across the lamp

      (ii) an ammeter to measure the current in the lamp.

   (b) The switch is now closed. A voltmeter connected across the lamp measures 1.5 V. What is the voltage across the resistor?

3. Calculate the missing voltage and current values in these circuits.

   (a)

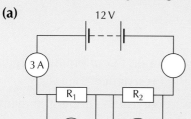

   (b)

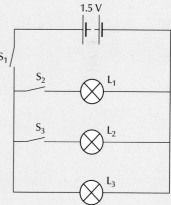

4. State one advantage that a parallel circuit has over a series circuit.

5. Look at the circuit shown on the right.

   (a) Which switches have to be closed for lamp L_3 to come on?

   (b) Which lamps are on when switches S_1 and S_3 are closed?

   (c) All switches are closed. What is the voltage across each lamp?

   (d) The lamps are now connected to the same cell in a series circuit with one switch.

      (i) How will the brightness of each lamp in the series circuit compare to the parallel circuit?

      (ii) Explain why the switch cannot be used to control each lamp independently, as was the case in the parallel circuit.

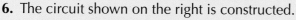

6. The circuit shown on the right is constructed.

   (a) State the readings on the two blank meters connected next to the bottom lamp.

   (b) What is the voltage supplied by the battery?

7. Kate is preparing to put the Christmas tree lights on the tree, and notices one is missing. The lamps will not work unless the missing one is replaced. Are the Christmas tree lights connected in series or parallel? Explain your answer.

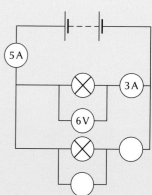

Resistors in circuits

Learning intentions

In this section you will:

- learn that resistance is the opposition to current in a circuit
- learn that the resistance of a resistor is measured in ohms (Ω)
- investigate the relationship between current, voltage and resistance.

📖 Word bank

- **Resistance**

Resistance is a measure of the difficulty for charges to flow through a conductor. It is measured in ohms (Ω).

All materials offer some opposition to current in a circuit. This opposition to current is called **resistance**.

Many circuits use resistors to control the current in them. If too much current passes through a laptop computer, for example, it could cause damage to the internal circuits. Resistors protect components from too high a current. The greater the resistance, the smaller the current. The resistance of a resistor is measured in ohms (Ω).

Types of resistor

There are two main types of resistors. Fixed resistors are usually very small and are manufactured to a specific value of resistance. Variable resistors come in all shapes and sizes and can be adjusted to give any resistance within a range.

(a) **(b)**

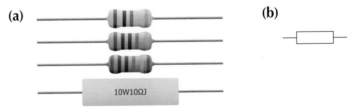

10W10ΩJ

Figure 4.19   *(a) Fixed resistors of different designs and values (b) Circuit symbol for a resistor.*

(a) **(b)**

Figure 4.20   *(a) Variable resistors of different designs and values (b) Circuit symbol for a variable resistor.*

Current, voltage and resistance

In the circuit in Figure 4.21 the current in the resistor is dependent on the 'push' provided by the voltage of the battery. The current is also dependent on the resistance of the resistor. The bigger the resistance, the lower the current. A relationship exists between all three of these properties of a simple circuit:

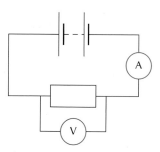

- when the resistance increases, the current decreases
- when the voltage increases, the current increases.

Figure 4.21   *A circuit designed to measure the voltage across and the current in a resistor.*

GO! ## Activity 4.10

1. Experiment In this activity, you will investigate how the current in a resistor and voltage across a resistor varies with the value of the resistor.

Read the plan carefully then follow the instructions, making sure you follow all of the safety rules as you would normally be expected to do when carrying out practical work.

You can work in small groups but you should record your own observations and complete the answers yourself.

You will need: a resistor, a voltmeter, an ammeter, a 6 V variable battery pack (four 1.5 V cells connected together) and connecting leads.

Method

(a) Set up the circuit shown.

(b) Copy the table below and use it to record your experimental findings:

Voltage across resistor (V)	Current in resistor (A)
1.5	
3.0	
4.5	
6.0	

(c) Connect the resistor to the 1.5 V cell only. Read the voltage across the resistor and the current through it using the voltmeter and ammeter, and record it in your table.

(d) Repeat the voltage and current measurements with the resistor connected to the 3 V, 4.5 V and 6 V cell combinations. Record the current readings in the table.

(e) What happens to the current in the resistor as the voltage across the resistor is increased?

(f) What happens to the current reading when the voltage across the resistor is doubled? (Hint: Check the readings for the current when the voltage is 1.5 V and 3 V, and at 3 V and 6 V.)

Activity 4.11

☺ Copy and complete the following paragraph about voltage using the words provided:

charges   resistance   doubles   opposition

voltage   twice   ohms

_ _ _ _ _ _ _ _ _ _ is a measure of the _ _ _ _ _ _ _ _ _ _ given to the flow of _ _ _ _ _ _ _ in a circuit. Resistance is measured in _ _ _ _. For a fixed resistor, if the _ _ _ _ _ _ _ across the resistor _ _ _ _ _ _ _, the current is also _ _ _ _ _ as large.

<table>
<tr><td>National 4</td></tr>
<tr><td>Curriculum level 4</td></tr>
<tr><td>Forces, electricity and waves: Electricity SCN 4-09a</td></tr>
</table>

Resistors and Ohm's Law

Learning intentions

In this section you will:

- investigate factors affecting resistance
- learn how to use the relationship $R_T = R_1 + R_2 + \ldots$ to calculate the total resistance in a series circuit
- investigate the relationship between current, voltage and resistance
- use the relationship between current, voltage and resistance to solve practical problems.

📖 Word bank

- **Conductor**

Conductors are materials that have a large amount of free electrons. These free electrons are easily pushed by a battery around a circuit to form a current.

The resistance of a wire depends on three main factors.

- The **material** the wire is made from. Some materials are better **conductors** than others. Silver is one of the best conductors with a low resistance. It has a large number of free electrons that increase its conductivity.

- The **thickness** of the wire. Similar to how a wide water pipe has a low resistance to the flow of water, a thick wire has low electrical resistance. It is easier to push the electrons through a thick wire, so there is less resistance. There are also more free electrons available to be pushed in a thicker wire, so the current is greater.

- The **length** of the wire. The longer the wire, the more resistance there will be. Resistance occurs as a result of the collisions between electrons and the atoms in the wire. There will be more collisions in a longer wire, so the resistance will be higher.

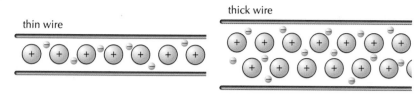

Figure 4.26 *Electrons can move through thick wires more easily than thin wires.*

GO! Activity 4.12

1. **Experiment** In this activity you will investigate how the current flowing in a wire changes with the length of the wire.

Read the plan carefully then follow the instructions, making sure you follow all of the safety rules as you would normally be expected to do when carrying out practical work.

You can work in small groups but you should record your own observations and complete the answers yourself.

You will need: different lengths of resistance wire (a range from approximately 5 cm to 50 cm), a 6 V battery, an ammeter, a switch and connecting leads.

Method

(a) Connect a length of resistance wire to the battery and the ammeter, as shown.

(b) Close the switch and use the ammeter to measure the current flowing in the circuit.

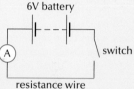

⚠ Do not leave the switch closed for a long time. Once you have recorded the current, open the switch to prevent overheating.

(c) Repeat the experiment for the different lengths of resistance wire. Make sure at least five different lengths of wire are used. Record the results in a table like this one.

Length of resistance wire (cm)	Current (A)

(d) Plot a line graph with the current on the vertical axis and the length of wire on the horizontal axis.

(e) Look at the graph you have drawn. How is the current affected by the length of wire? What conclusion can you make about how the length of wire affects its resistance?

(f) What could you do to increase the reliability of your results?

(g) Imagine you conducted another experiment to determine how the thickness of the wire affected the current flowing through it. What would you have to do to ensure it was a fair test?

Figure 4.23 *A multimeter.*

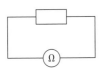

Figure 4.24 *An ohmmeter connected to a resistor. The resistor should not be connected to a battery when the ohmmeter is measuring its resistance.*

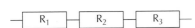

Figure 4.25 *Three resistors connected in series.*

Resistors in series circuits

Current and voltage can be measured using a device called a **multimeter**.

A multimeter can also measure the resistance of a resistor. Devices specifically designed for measuring electrical resistance are called ohmmeters. The multimeter is behaving like an ohmmeter when its dial is turned to measure resistance.

Adding resistors in series

In Activity 4.12 each wire had a different resistance based on its length. The longer the wire, the greater the resistance. A simple way to make a larger resistor from a number of smaller ones is to connect them together, one after another. This is known as connecting resistors in series.

The total resistance of a number of resistors connected together in series is found by adding the resistance of each resistor together. This can be summarised in an equation:

$$R_T = R_1 + R_2 + R_3$$

where:

R_T is the total resistance

R_1, R_2 and R_3 are the separate resistors.

By adding resistors together in series, the total resistance increases and the current in the resistors will decrease.

 Activity 4.13

 1. Experiment In this activity you will investigate the relationship for adding resistors in series, using resistance wire. This experiment can be completed with three fixed resistors of different sizes if resistance wire is not available.

Read the plan carefully then follow the instructions, making sure you follow all of the safety rules as you would normally be expected to do when carrying out practical work.

You can work in small groups but you should record your own observations and complete the answers yourself.

⚠ Resistance wire can have sharp ends. Be careful when you handle it in this experiment.

You will need: an ohmmeter, three lengths of resistance wire, some crocodile clips and four connecting leads.

Method

(a) Using the ohmmeter and two connecting leads with crocodile clips, measure the resistance of each length of resistance wire (as shown). Record the results in a table like the one shown.

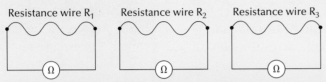

Resistance wire	Resistance of resistance wire (Ω)
R_1	
R_2	
R_3	
R_1, R_2 and R_3 connected together in series	

(b) Connect each length of resistance wire together using crocodile clips and connecting leads. Use the ohmmeter to measure the total resistance of the wires when connected in series, by connecting it across the ends of the wires as shown.

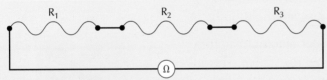

(c) How does the total resistance R_T compare to the resistance of the three wires when added together ($R_1 + R_2 + R_3$)?

Example 4.3

Three resistors are connected in series as shown.

| 50Ω | 10Ω | 33Ω |

Calculate the total resistance between the terminals.

$R_1 = 50 \ \Omega$

$R_2 = 10 \ \Omega$ ●————————(Write out what you know from the question.)

$R_3 = 33 \ \Omega$

$R_T = ?$

$R_T = R_1 + R_2 + R_3$ ●————————(Write out the equation.)

$R_T = 50 + 10 + 33$ ●————————(Substitute in what you know.)

$R_T = 93 \ \Omega$ ●————————(Don't forget to write the units in your answer.)

| N3 | L3 | **N4** | L4 | **83** |

✴ Physics in action: Resistance creates heat!

When there is current in a wire with a high resistance, it creates heat. The electrons collide with the stationary atoms in the metal and some of their kinetic energy is transferred into heat energy.

The heating coil in a toaster or electric heater is made of high-resistance wire. When current passes through the wire it heats up and glows red hot.

Figure 4.26 *The glowing resistance wires inside a toaster.*

The filament wire in an incandescent light bulb is actually a resistor. It is made of very thin wire made of tungsten, which has a very high resistance. The filament wire resists the current in the bulb. The wire heats up to a temperature that produces light.

Incandescent light bulbs are highly inefficient, producing much more heat than light. In 2009, the European Union set up regulations to begin phasing out the sale of incandescent light bulbs. Most homeowners are now replacing them with more energy-efficient light bulbs.

Figure 4.27 *A tungsten filament bulb.*

Transmission lines are used to transfer electricity across large distances. They have a very low resistance. The voltage of the cables is around 400 000V, which enables the current in the cables to remain low, even when the resistance is low. If the cables had high resistance, the energy wasted due to heat being transferred to the surroundings would be significant.

Figure 4.28 *Despite being an eyesore, transmission lines allow electricity to travel long distances without wasting too much energy as heat in the wires.*

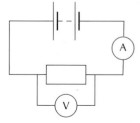

Figure 4.29 *A circuit designed to measure the voltage across and the current in a resistor.*

Ohm's Law

The relationship between current, voltage and resistance can be investigated using the circuit shown in Figure 4.29.

By measuring the voltage across the resistor using the voltmeter, and the current through the resistor using the ammeter, the resistance can be determined.

The resistance of the resistor is calculated using the relationship shown.

| N3 | L3 | N4 | L4 |

$$\text{resistance} = \frac{\text{voltage across resistor}}{\text{current through resistor}}$$

This can be written using symbols:

$$R = \frac{V}{I}$$

This relationship is known as **Ohm's Law**. It is normally written in the form:

$$V = IR$$

That is:

voltage across resistor = current through resistor × resistance

An ohmmeter does not measure resistance directly. It determines the voltage across the resistor and the current through it, and calculates the resistance using Ohm's Law.

GO! Activity 4.14

1. Experiment In this activity you will investigate and confirm Ohm's Law.
Read the plan carefully then follow the instructions, making sure you follow all of the safety rules as you would normally be expected to do when carrying out practical work.

You can work in small groups but you should record your own observations and complete the answers yourself.

You will need: an ohmmeter, four resistors of different sizes, a voltmeter, an ammeter, a 6V battery and five connecting leads.

Method

(a) Using the ohmmeter, measure the resistance of each of the four resistors and record the values in a table like the one shown below.

Resistor	Resistance (Ω)	Voltage across resistor (V)	Current in resistor (A)	Voltage/current
A				
B				
C				
D				

(b) Set up the circuit using resistor A.

(c) Measure the voltage across the resistor and the current through it using the voltmeter and ammeter, and record it in your table.

(d) Repeat the voltage and current measurements for the other three resistors.

(e) For each resistor, divide the voltage by the current and record the calculated value in the last column of the table.

(f) How do the values in the second column (Resistance) compare with the last column (Voltage/current)?

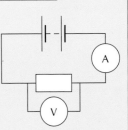

Example 4.4

The current in a resistor is 0.2 A. The resistance of the resistor is $50\,\Omega$.
Calculate the voltage across the resistor.

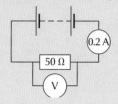

$I = 0.2\,A$

$R = 50\,\Omega$ •————————(Write out what you know from the question.)

$V = ?$

$V = IR$ •————————(Write out the equation.)

$V = 0.2 \times 50$ •————————(Substitute in what you know.)

$V = 10V$ •————————(Don't forget to write the units in your answer.)

Example 4.5

The voltage across a lamp is 6 V when the current in the lamp is 1.2 A.
Calculate the resistance of the lamp.

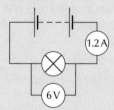

$V = 6V$

$I = 1.2\,A$ •————————(Write out what you know from the question.)

$R = ?$

$V = IR$ •————————(Write out the equation.)

$6 = 1.2 \times R$ •————————(Substitute in what you know and solve for R.)

$R = \dfrac{6}{1.2} = 5\Omega$ •————————(Don't forget to write the units in your answer.)

| N3 | L3 | **N4** | L4 |

GO! Activity 4.15

☺ Use this activity to review what you have learned about resistors in circuits.

1. State three factors that affect the resistance of a resistor.

2. A variable resistor is used to control the current in a lamp. The variable resistor is adjusted in such a way as to increase the length of wire that the current has to travel through.

   (a) State the effect of this adjustment on the resistance of the variable resistor.

   (b) What happens to the brightness of the lamp? Explain your answer.

3. There is a current of 2 A in a resistor of 10 Ω. Calculate the voltage across the resistor.

4. A lamp is connected to a 6 V battery. There is a current of 0.5 A in the lamp. What is the resistance of the lamp?

5. The rear reverse light on a car is connected to the 12 V battery. The reverse light has a resistance of 3 Ω. What is the current in the light when the car is reversing?

6. Three resistors are connected together in series. The resistance of the resistors is 2 Ω, 15 Ω and 60 Ω. Calculate the total resistance in the circuit.

7. The table below compares the resistance of different materials.

Material	Resistance value (Ω x 10^{-8} m)
Copper	1.72
Silver	1.59
Tungsten	5.6
PVC	2 000 000

   (a) Use the information in the table to explain why copper wires are insulated with a PVC jacket.

   (b) Suggest why copper is used for electrical cabling instead of silver, even though silver has a lower resistance.

   (c) Explain why a tungsten filament wire heats up when there is a current in it but the copper wires connecting it to a battery do not.

8. A lamp is found on the floor of a physics classroom. The markings have been worn off the glass. Draw a circuit that could be used to determine the resistance of the lamp, and explain what steps must be taken to determine its resistance.

Learning checklist

After reading this chapter and completing the activities, I can:

N3 L3 N4 L4

- describe current as the movement of electrical charges around a circuit. **Activity 4.1** ○ ○ ○

- state that current is measured in amps. **Activity 4.1** ○ ○ ○

- state that current is measured with an ammeter. **Activity 4.1** ○ ○ ○

- describe voltage as a measure of the energy given to charges in a circuit. **Activity 4.2** ○ ○ ○

- state that voltage is measured in volts. **Activity 4.2** ○ ○ ○

- state that voltage is measured with a voltmeter. **Activity 4.2** ○ ○ ○

- explain the features of series and parallel circuits and differences between them. **Activity 4.3, Activity 4.4** ○ ○ ○

- state that current is measured in series with a component. **Activity 4.5, Activity 4.7, Activity 4.9 Q2(a)** ○ ○ ○

- state that the current in a series circuit is the same at all points in the circuit. **Activity 4.5, Activity 4.9** ○ ○ ○

- state that voltage is measured in parallel across a component. **Activity 4.6, Activity 4.8, Activity 4.9 Q2(a)** ○ ○ ○

- state that the sum of the voltages across the components in a series circuit is equal to the supply voltage. **Activity 4.6, Activity 4.9** ○ ○ ○

- state that the current leaving the supply is equal to the sum of the current through each branch in a parallel circuit. **Activity 4.7, Activity 4.9** ○ ○ ○

- state that the voltage across the supply is equal to the voltage across each branch in a parallel circuit. **Activity 4.8, Activity 4.9** ○ ○ ○

N3 L3 N4 L4

- investigate the relationship between current, voltage and resistance and state that when the voltage across the resistor doubles, the current through the resistor doubles. **Activity 4.10, Activity 4.11**   ○ ○ ○

- state that the resistance of a resistor is measured in ohms (Ω). **Activity 4.11**   ○ ○ ○

- state that resistance is the measure of opposition to current in a circuit. **Activity 4.11**   ○ ○ ○

- explain factors that affect resistance. **Activity 4.12, Activity 4.15 Q1, Q2**   ○ ○ ○

- calculate the total resistance of a number of resistors connected in a series circuit. **Activity 4.13, Activity 4.15 Q6**   ○ ○ ○

- state that the relationship between current, voltage and resistance for a fixed resistor is $R = V/I$. **Activity 4.14**   ○ ○ ○

- use the relationship between current, voltage and resistance to solve problems. **Activity 4.15**   ○ ○ ○

5 Batteries and cells

This chapter includes coverage of:

N3 Electricity • Forces, electricity and waves SCN 3-10a • Forces, electricity and waves SCN 4-10b • Topical science SCN 4-20a

You should already know:

- how to build simple chemical cells using readily available materials to provide power for appliances.

Curriculum level 3

Forces, electricity and waves: Electricity SCN 3-10a

Curriculum level 4

Forces, electricity and waves: Electricity SCN 4-10b; Topical science SCN 4-20a

Cells and batteries

Learning intentions

In this section you will:

- explain the difference between a cell and a battery
- describe the function of a simple chemical cell
- investigate the latest developments in chemical cell technology
- describe the impact of chemical cells in modern life.

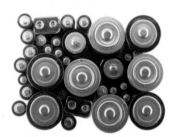

Figure 5.1 *Chemical cells come in all shapes and sizes and are used in everything from small watches to large vehicles.*

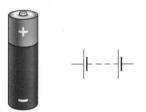

Figure 5.2 *A cell and its circuit symbol.*

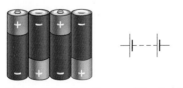

Figure 5.3 *A battery and its circuit symbol.*

Many electrical devices today use chemical cells to convert chemical energy into electrical energy.

A **battery** is a series of chemical cells connected together.

By connecting cells together in this way, the overall voltage produced increases.

A **dry cell** is the most common type of chemical cell.

The first dry cells were developed by a German scientist called Carl Gassner in 1886. They were called zinc–carbon batteries, and are still made today. They have three main parts:

- a positive electrode made of carbon (called the cathode)
- a negative electrode made of zinc (called the anode)
- an electrolyte (a paste made of ammonium chloride).

A complex chemical reaction takes place when the cell is connected in a complete circuit like the one shown in Figure 5.5.

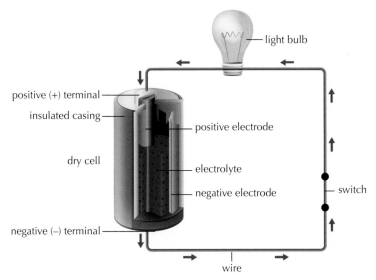

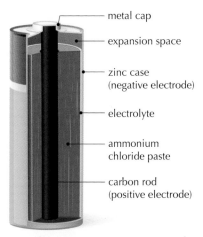

Figure 5.4 *A zinc–carbon cell.*

Figure 5.5 *The cell pushes electrons around the circuit.*

Electrons build up on the zinc anode (negative terminal), which pushes electrons in the wire around the circuit to the positive cathode. A **voltage** is generated between the anode and the cathode. This movement of electrons forms an **electric current**, which transfers energy to the components in the circuit.

Applications of cells

Advances in cell technologies are having a significant impact on modern life.

Smartphones and tablets contain lithium-ion batteries. They are rechargeable, meaning the chemical reaction in the battery is reversible. Lithium-ion batteries were invented at Oxford University in the 1970s. About five billion are made every year, mostly in China.

The miniaturisation of batteries in modern smartphones has revolutionised how we communicate and manage our lives. Scientists and engineers are constantly researching new materials to allow batteries to become smaller, charge faster and last longer.

Modern electric cars such as the Tesla Model S have over 7000 lithium-ion cells. These cells enable the car to travel over 260 miles on a single charge, making battery powered electric cars a viable transport solution for many road users.

Large-scale batteries are used to store electrical energy when production from renewable energy sources exceeds demand. This energy can be returned to the national electric grid when production falls below demand. As Scotland seeks to increase its reliance on renewable energies in the future, the development and use of batteries for energy storage is likely to rise.

📖 Word bank

• **Electrode**

An electrode is a conductor that passes electric current from one substance to another. The electrodes in a cell allow charges to pass between the wires connected to the cell and the electrolyte paste inside the cell.

❓ Did you know ...?

Alkaline cells have all but replaced zinc–carbon cells. They are almost the same as zinc–carbon cells except that the electrolyte paste is made of potassium hydroxide. They tend to last much longer than zinc–carbon cells, but they are prone to leaking.

Figure 5.6 *A pair of leaked alkaline cells.*

🔬 Make the link

Renewable energies are described in Chapter 1.

| N3 | **L3** | N4 | **L4** |

Figure 5.7 *Old mobile phones had large batteries making them impractical.*

Figure 5.8 *Tesla Model S being charged at a Tesla Supercharger station.*

GO! Activity 5.1

☺ **1.** State the energy change that takes place in a chemical cell.

2. Describe the three main parts of a zinc–carbon cell.

3. Explain how a voltage can be produced by a zinc–carbon cell.

4. For one of the following applications of chemical cells:

   (a) miniaturisation of cells used in smartphones

   (b) high-energy cells used in electric vehicles

   (c) batteries used for large-scale electricity storage,

      (i) research the latest developments for that application

      (ii) describe the impact the application has had on modern life.

National 3
Curriculum level 3
Forces, electricity and waves: Electricity SCN 3-10a

Figure 5.9 *Electrons move through the solution from the copper plate to the zinc plate. The current is measured with an ammeter.*

Figure 5.10 *A lemon battery.*

Fruit cells

Learning intentions

In this section you will:

• investigate the operation of a simple chemical cell

• investigate factors that affect the voltage produced by a chemical cell.

When copper and zinc in contact with each other, zinc will naturally give away some electrons to the copper. The movement of charges stops quite quickly, but if the metals are connected in a circuit and suspended in an acidic solution (Figure 5.9), the flow of charge is steady. This arrangement is the basis of a simple chemical cell.

The size of the current and voltage between the zinc and copper depends on the acidic solution used.

GO! Activity 5.2

1. Experiment In this activity you will investigate how the voltage varies when copper and zinc are placed into different fruit, vegetables or even fizzy drinks.

Read the plan carefully then follow the instructions, making sure you follow all of the safety rules as you would normally be expected to do when carrying out practical work.

You can work in small groups but you should record your own observations and complete the answers yourself.

You will need: the apparatus shown in the diagram.

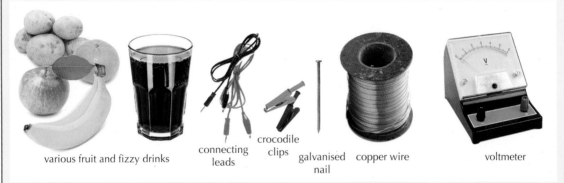

various fruit and fizzy drinks   connecting leads   crocodile clips   galvanised nail   copper wire   voltmeter

Method

(a) Insert the copper wire and zinc nail into the food sample, and attach the two connecting leads with crocodile clips to the copper and zinc. Connect the other ends of the leads to the voltmeter.

(b) Record the value of the voltage in the table.

(c) By inserting the copper wire and zinc nail into the other food samples, record the voltage for each sample in a table like the one below. Remember to include the correct unit.

Fruit/vegetables/fizzy drink	Voltage produced (V)

(d) From your observations, write down a conclusion about what effect changing the fruit had on the voltage produced.

(e) Can you think of any other ways to achieve a higher voltage from your 'best' fruit? If you have time, test your idea.

🔍 Hint

Think about adding more copper wire and nails to the fruit. What do you think would happen if you connected two fruit together? Perhaps investigate using different metals for the electrodes?

Photocells

Figure 5.11 *A photocell transfers light energy into electrical energy for an outdoor garden lamp.*

Learning intention

In this section you will:

- investigate factors that affect the voltage produced by a photocell.

A photovoltaic cell, or **photocell**, transfers light energy into electricity. Photocells are often used to power calculators. Many homes have photocells on the roof to convert sunlight directly into electricity. Large panels of photocells are also used to provide power for satellites in orbit around the Earth.

The output of a photocell depends on two things:

- the surface area of the photocell
- the intensity of the light shining on the photocell.

 Activity 5.3

1. Experiment In this activity, you will investigate the factors that affect the output voltage of a photocell.

Read the plan carefully then follow the instructions, making sure you follow all of the safety rules as you would normally be expected to do when carrying out practical work.

You can work in small groups but you should record your own observations and complete the answers yourself.

You will need: a photocell, a voltmeter, a lamp or torch, a metre stick and some paper.

Method

1. Connect the photocell to the voltmeter, and set it up so it is facing the lamp.

2. Turn on the lamp and record the output voltage of the photocell when it is positioned at a range of distances (between 10 cm and 100 cm) from the lamp. Record the results in a table like the one below.

Arrangement of the photocell at varying distances from a lamp.

Distance between photocell and lamp (cm)	Voltage produced (V)
10	
20	
30	
...	
100	

3. Describe how the voltage output from the photocell is affected by the distance the lamp is from the photocell.

4. Now position the lamp at 20 cm from the photocell. Using paper, cover a 1 cm width of the photocell and take a reading of the voltage output. Record the output voltage of the photocell in a separate table.

5. Increase the amount of photocell covered by a further 1 cm width, and record the reading on the voltmeter. Repeat for 1 cm width increments until the photocell is completely covered, noting the voltage reading each time.

6. Describe how the voltage output from the photocell is affected by the area of the cell that is exposed to the light.

Learning checklist

After reading this chapter and completing the activities, I can:

N3 L3 N4 L4

- use the terms electrode and electrolyte correctly and explain their function in a simple chemical cell. **(Activity 5.1 Q1–Q3)** ○ ○ ○

- describe developments in chemical cell technology and evaluate their impact on society. **(Activity 5.1 Q4)** ○ ○ ○

- design a simple chemical cell and use it to investigate the factors which affect the voltage produced. **(Activity 5.2)** ○ ○ ○

- investigate factors that affect the voltage output from a photocell. **(Activity 5.3)** ○ ○ ○

6 Practical electricity and safety

National 3

Curriculum level 3

Forces, electricity and waves: Electricity SCN 3-09a

National 4

Electricity in the home

Learning intentions

In this section you will:

- identify examples of parallel circuits in everyday applications
- investigate electrical circuits used in the home
- investigate simple circuits to solve practical problems in the home
- identify factors affecting electrical safety including double insulation, the three-pinned plug and human conductivity
- learn the circuit symbol, function and application of a fuse as a safety device
- learn about government standards and consumer advice.

Series and parallel circuits are used in all types of devices in and around our homes. Without them, many of the modern conveniences of everyday life would not exist.

Circuits in the home

Every plug socket in a home is connected to a parallel circuit known as a **ring main**. Since it is connected in parallel, every plug socket has the same 230 V applied to it. Each appliance can also be turned on and off without affecting the other appliances.

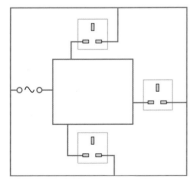

Figure 6.1 *Sockets connected in parallel in a household ring main circuit.*

Lighting circuits are also wired in parallel. This ensures that each lamp receives the same voltage, and if one lamp is turned off or breaks, the others can still operate normally. Dimmer switches are often connected in series with room lamps. Turning the dial adjusts the resistance of a variable resistor. When it is turned clockwise, the resistance decreases and there is more current in the lamp, making it brighter.

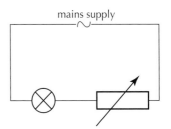

Figure 6.2   *Series circuit for a dimmer switch.*

Solving practical problems with simple circuits

With some creative thinking, simple circuits can be designed to solve practical problems in and around the home.

Continuity testers

Continuity testers can be used to test whether a material conducts electricity, or if there is a break in a circuit. The simplest continuity tester is a series circuit that causes a light to light up when current passes through the material being tested.

If current flows in the circuit, the circuit is complete and the light will switch on. If there is no current, the light won't switch on. This indicates that there is a break in the circuit. Many continuity testers will also make a sound if the circuit is complete. For example, multimeters used in classrooms have a continuity tester that will buzz when the circuit under test is working correctly.

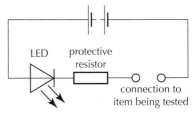

Figure 6.3   *Circuit diagram for continuity tester.*

Stair lighting

Light switches located at the top and bottom of a flight of stairs can control the current in a lamp. Two or more two-way switches are connected in series to allow either switch to turn on the lamp.

Figure 6.4   *Circuits can be tested using the probes of a multimeter.*

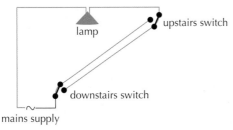

Figure 6.5   *The lamp is off. Closing either switch will turn on the lamp.*

Figure 6.6 *A hairdryer with switches for heater and fan motor.*

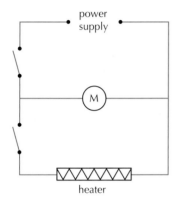

Figure 6.7 *Parallel circuit for a hairdryer.*

Figure 6.8 *Car sidelight on without headlight being on.*

Hairdryers

Hairdryers typically have cool and hot settings. The circuit diagram in Figure 6.7 shows how the motor can be switched on without the heater, producing cool air. Only by closing the second switch will hot air be produced. The heater cannot be turned on without the fan running.

Car lighting

Although motor cars are becoming increasingly sophisticated, all cars for many years have used parallel lighting circuits.

The two sidelights are connected in parallel with the two headlights. Switch 1 in Figure 6.9 allows the sidelights to turn on. Switch 2 controls the headlights. Current cannot get to the headlights unless switch 1 is closed first. This means that cars wired in this way cannot have their headlights on without the sidelights being switched on first.

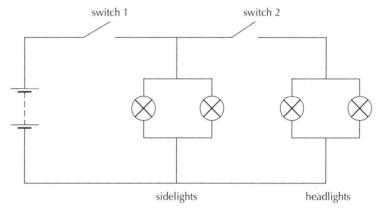

Figure 6.9 *Parallel circuit for side and headlights.*

Advantages of parallel circuits in everyday applications

Parallel circuits are used for many applications in and around the home, such as the ring mains connecting the sockets, and the lighting circuits. Devices such as hairdryers, ovens and dishwashers all use parallel circuits. Lighting and heating circuits in vehicles use parallel circuits.

The two main reasons for using parallel circuits are:

- Each component in the appliance has the same voltage.

- If one component stops working, the others will still operate.

GO! Activity 6.1

1. A pencil sharpened at both ends is connected into a circuit with a lamp and a cell. The lamp lights up.

   Does pencil lead conduct?

   (a) What does this tell you about pencil lead?

   (b) What is the name given to the type of circuit used to test whether the pencil conducts electricity?

2. Stair lighting can often be controlled by more than one switch.

   (a) Explain why it is helpful to be able to control the lighting from either upstairs or downstairs.

   (b) Use a diagram to explain how two-way switches are used to enable more than one switch to control a lamp.

3. Electrical circuits in the home are usually wired in parallel. State two reasons why it is necessary to wire houses with parallel circuits.

4. A manufacturer of underfloor heating mats creates two designs for an electrical circuit for a new heating mat. Each design has two heating coils. The designs are shown below.

   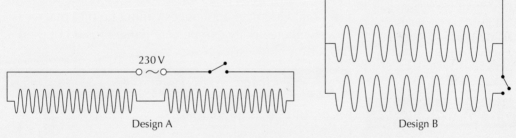

   Design A   Design B

   The manufacturer chooses to use Design B for its new heating mat.

   (a) State what type of circuit Design B is.

   (b) Explain what would happen if one of the heating coils in Design A stopped functioning.

   (c) Design B has two heating settings. Explain how the switches can be used to operate the two heat settings.

5. A data projector uses a bright lamp designed to operate at mains voltage (230V). The projector also requires a fan motor to ensure the lamp doesn't overheat. Design a parallel circuit that would allow the lamp to light, but only when the fan motor is switched on first. (To get started, look at the circuit diagram for the hairdryer Figure 6.7.)

   (continued)

6. A circuit diagram for a car electrical system is shown below.

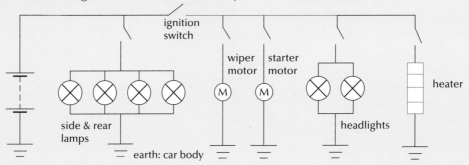

(a) State what components require the ignition to be switched on before they will operate.

(b) (i) Explain why the side and rear lamps are connected in parallel.

   (ii) Give a reason why the side and rear lights can be operated without the ignition being switched on.

Electrical safety in the home

Electricity has transformed the way we live. 100 years ago only 10% of the UK had an electricity supply. The appliances were mostly lights, irons and radios. Today, there are few household tasks that electrical appliances cannot help to make easier. However, electricity can be very dangerous if not used properly.

📖 Word bank

• **Conductivity**

Conductivity is the degree to which a material conducts electricity. A material that has high conductivity generally has a low resistance.

❓ Did you know ...?

When people tell lies their hearts start to race. This drives up blood pressure and sweating occurs more easily. Sweat increases the conductivity of the skin. Lie detectors measure the resistance of the skin and can determine if someone is lying due to the increased conductivity of their skin.

Figure 6.11 *A lie detector measures an increase in human conductivity because we sweat more when lying.*

❓ Did you know ...?

In the UK, electricity is the major cause of accidental fires. Government statistics show that electricity causes more than 20 000 fires a year in the UK. On average, 70 people die each year and 350 000 are seriously injured due to electrical accidents in the home.

Figure 6.10 *Overloading mains sockets increases the fire risk due to electricity.*

Human conductivity

The human body has a naturally high resistance to electric current. For comparison, the copper wiring in the mains circuit of a house has a resistance of around a few ohms, but the human body's resistance is about 1000 ohms. However, water significantly increases **conductivity**.

Pure water is not very conductive, but the minerals that are dissolved in water make it more conductive. If you touch an electrical appliance with wet hands, you are more likely to get an electric shock.

Activity 6.2

1. Experiment In this activity you will compare the resistance of the skin with dry and wet hands.

Read the plan carefully then follow the instructions, making sure you follow all of the safety rules as you would normally be expected to do when carrying out practical work.

You can work in small groups but you should record your own observations and complete the answers yourself.

You will need: a multimeter and two connecting leads.

Method

(a) Connect the leads to the multimeter and set the dial to measure up to 2000 Ω.

(b) Touch the ends of each lead on to the skin on your hand, and take a note of the resistance.

A multimeter can be set to measure resistance by turning the dial to the ohm symbol (Ω).

(c) Wet your hands, and repeat step **(b)** to measure the resistance with wet hands.

(d) Is there a noticeable change in your skin's resistance when it is wet? Why is it important to avoid using electrical appliances with wet hands?

Appliances and safety

Appliances that are used in our homes are designed with a number of safety features in mind.

Insulated flexible plastic cable containing three inner wires

Electricity supply can be turned off easily

Mains plug designed with safety in mind

Figure 6.12 *Appliances are designed with safety in mind.*

Insulated plastic cabling

The mains cable contains two or three wires, made of copper, which is a good conductor. All these electric wires are covered in plastic, which is a good insulator. The inner wires are colour-coded so the purpose of each wire can be easily identified.

Isolating switch

Most appliances have their own on/off switch. All appliances can also be turned off with the mains switch on the wall socket.

Table 6.1 *Properties of the wires found in a mains cable.*

Name of wire	Colour	Purpose
Live	Brown	These two wires provide a route for charges to flow to the appliance from the mains supply
Neutral	Blue	
Earth	Yellow and green	A safety wire which provides a route for charges to flow through if a fault occurs in the appliance

The mains plug

The UK mains plug is a marvel of British engineering and design.

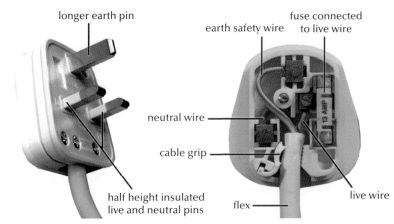

Figure 6.13 *The British plug is one of the safest plug designs in the world.*

The live wire (brown) and neutral wire (blue) are connected to the two shorter pins. The live wire is also connected to a **fuse**. The fuse is a small tube which contains a piece of wire which is designed to melt if the current gets too high. If the fuse melts, then the circuit is broken and no more charges can flow through the live wire to the appliance.

If a fault develops in an appliance, there may be too much current in the wires to the appliance. High current can cause the wires to heat up. Fuses protect both the appliance and the wires in the walls of the house from overheating and catching fire.

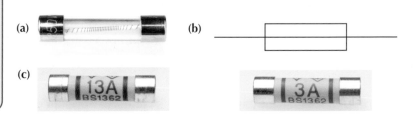

Figure 6.14 *(a) The fuse wire can be seen inside the glass body of the fuse. (b) The circuit symbol for a fuse. (c) Two common fuse values in the UK.*

Activity 6.3

1. In this activity you will investigate the current required to melt a fuse made of steel wool. Read the plan carefully then follow the instructions, making sure you follow all of the safety rules as you would normally be expected to do when carrying out practical work.

You can work in small groups but you should record your own observations and complete the answers yourself.

⚠ The steel wool will melt in this experiment, so safety goggles must be worn. Also, **do not** touch the steel wool once it is connected to the circuit.

You will need: three 1.5 V cells, seven connecting leads, a component holder with crocodile clips, a small amount of steel wool, a heat-resistant mat, a 1.5 V lamp, a variable resistor, a switch, and an ammeter.

Method

(a) Set up the circuit as shown, twisting a few strands of the steel wool together and attaching them between crocodile clips.

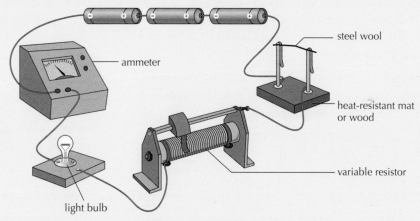

Circuit set-up to run current through steel wool.

(b) Set the resistor to its highest resistance and complete the circuit using crocodile clips. The lamp should light. What is the reading on the ammeter?

(c) Decrease the resistance of the resistor slowly. Observe what happens to the current reading on the ammeter. Can you explain why this is happening?

(d) Continue to decrease the resistance of the variable resistor. Eventually the steel wool will melt. Record the ammeter reading at which the steel wool melts.

(e) Can you explain what happens to the ammeter reading when the steel wool has melted?

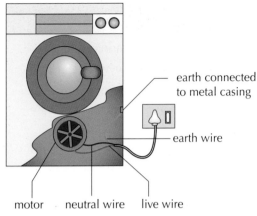

Figure 6.16 *The earth wire is connected to the metal casing inside a washing machine.*

motor   neutral wire   live wire

earth connected to metal casing

earth wire

The earth wire is coloured green and yellow. If an appliance has a metal casing, the earth wire is connected to the metal casing.

A fault may develop in which the live wire touches the metal casing of the appliance. If this happens then the appliance will become live, and anyone who touched it would be electrocuted. However, because the earth wire is attached to the casing, it provides a low resistance path for the current directly into the earth. This results in a high current in the earth wire. This high current melts the fuse, cutting off the electricity supply from the appliance and making it safe.

 Activity 6.4

1. Experiment The pins of a UK plug are designed with safety in mind. In this activity you will explore the safety features of the pins of a plug.

Read the plan carefully then follow the instructions, making sure you follow all of the safety rules as you would normally be expected to do when carrying out practical work.

You can work in small groups but you should record your own observations and complete the answers yourself.

You will need: an extension cable.

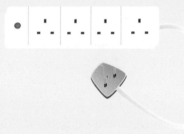

Extension cable.

⚠️ **Do not plug the extension cable into the mains socket. This experiment can be done without using a mains socket.**

Method

   (a) Take the plug end of the extension cable and insert the earth pin half way into one of the sockets on the extension cable.

   (b) What do you notice happens to the socket ports for the live and neutral terminals?

   (c) Insert the live and neutral pins in half way.

   (d) Explain the purpose of the insulation around the live and neutral pin.

   2. Draw a diagram of the UK plug and explain on it the two key safety features that the pins of the plug enable.

Double insulation

Many appliances in the home do not have a metal casing, but instead have a plastic exterior.

In these appliances, no earth wire is required. The exterior of the appliance is plastic and cannot become live. If a fault were to develop inside the appliance, there is no path for charges to flow to the user. This makes the appliance safe.

Appliances like these are called **double insulated**. Any appliance like this will have an information plate stating it is double insulated.

Figure 6.17 *These appliances all have plastic exteriors.*

Government standards

If you look closely at a mains plug, you will see British standard codes on the plug and on the fuse.

These codes refer to standards set out by the government to ensure electrical equipment is manufactured to meet high requirements for safety. The government has also produced regulations and advice to ensure that electrical equipment is handled safely at home, in schools and in places of work.

> ### 📢 Activity 6.5
>
> ☺ 1. Explain why it is dangerous to handle electrical appliances with wet hands.
>
> 2. Describe an experiment that could be carried out to test the difference in conductivity of wet and dry hands.
>
> 3. The British government produce safety codes for the production and use of electrical appliances. Use the government's Health and Safety Executive website (http://www.hse.gov.uk/electricity/standards.htm) to determine the purpose of the following safety codes:
>
>   **(a)** BS 1362   **(b)** BS EN 60335   **(c)** BS 5839.

> ### 📖 Word bank
>
> • **Double insulated**
>
> An electrical appliance is double insulated if it is designed so that it does not require an earth wire. The outer casing of double-insulated appliances is made of plastic. Two layers of insulation protect the user.

Figure 6.18 *The double insulated symbol on a TV.*

Figure 6.19 *The British safety code BS 1363 on the three-pin mains plug.*

> ### 🔄 Keep up to date!
>
> Advice from the government on electrical safety can be found at www.hse.gov.uk. Search online to explore the guidance on electricity at this site. Try to find out about the different safety precautions that can be implemented in different areas of work.

(continued)

4. This figure shows a diagram of the inside of a mains plug.

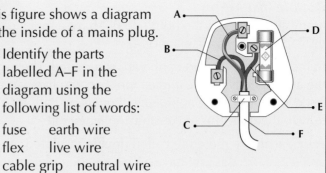

(a) Identify the parts labelled A–F in the diagram using the following list of words:

fuse   earth wire
flex   live wire
cable grip   neutral wire

(b) State the purpose of the fuse in the plug.

(c) Some appliances do not require an earth wire. These appliances are double insulated.

(i) Draw the symbol that identifies an appliance that is double insulated.

(ii) State the main difference between a double-insulated appliance and one that requires an earth wire.

Learning checklist

After reading this chapter and completing the activities, I can:

N3 **L3** **N4** L4

• explain the advantages of parallel circuits in an everyday situation. **Activity 6.1 Q3–Q6**	◯ ◯ ◯
• explain how electrical circuits used in the home are designed. **Activity 6.1**	◯ ◯ ◯
• give examples of electrical circuits that solve three practical problems; lighting stairways, providing underfloor heating, and operating car lights. **Activity 6.1 Q2, Q4, Q5**	◯ ◯ ◯
• explain the circuit diagrams of hairdryers and vehicle electrical systems. **Activity 6.1 Q5, Q6**	◯ ◯ ◯
• describe how plugs are designed to improve electrical safety in the home. **Activity 6.2, Activity 6.3, Activity 6.4, Activity 6.5**	◯ ◯ ◯
• describe the purpose of a fuse as a safety device. **Activity 6.5 Q3**	◯ ◯ ◯
• explain government standards and consumer advice for working with electrical appliances. **Activity 6.5 Q4**	◯ ◯ ◯

7 Electronic circuits

This chapter includes coverage of:

N4 Electricity and energy • Forces, electricity and waves
SCN 4-09b • Forces, electricity and waves SCN 4-09c

You should already know:

- that electrical current flows in circuits
- how to draw simple circuit diagrams.

Electrical devices

Learning intentions

In this section you will:

- learn that an electronic system has three main parts: input, process and output
- name and identify the function and application of standard electrical and electronic components
- identify analogue and digital input and output devices.

National 4

Curriculum level 4

Forces, electricity and waves: Electricity
SCN 4-09b

An **electronic system** can be thought of as any collection of electronic devices working together for a specific purpose. The devices used in an electronic system will vary for each purpose. A car stereo is an example of an electronic system.

Although modern car stereos can do many complex tasks, one key function of playing a radio station can be broken down into three key stages called sub-systems. These sub-systems are input, process and output. The sub-systems for the car stereo playing a radio station are shown in Table 7.1.

Table 7.1 *Input-process-output for a car radio.*

Sub-system	Device	Description
Input	Aerial	Radio waves picked up by the aerial are changed into an electrical signal.
Process	Amplifier	The electrical signal is amplified to make it stronger by a processing device called an amplifier.
Output	Loudspeaker	This amplified electrical signal is converted into sound with a loudspeaker.

📖 Word bank

- **Electronic system**
An electronic system is a set of electronic devices working together for a specific function.

Figure 7.1 *A car stereo has a number of devices that work together in an electronic system.*

Typically, both the input and output devices convert energy from one form to another. In the car stereo example, the aerial converts the energy of the radio signal into electrical energy and the loudspeaker converts electrical energy into sound energy. The processing device always alters the input in some way that can be used by the output device. In the car stereo example, without the amplifier, the signal from the radio would not be strong enough to power the loudspeaker.

All electronic systems can be represented by a block diagram. Each block represents a device, and the arrows show how they are connected. The block diagram for a car stereo is shown in Figure 7.2.

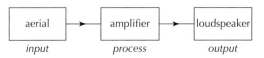

aerial	amplifier	loudspeaker
input	*process*	*output*

Figure 7.2 *A block diagram for a car stereo.*

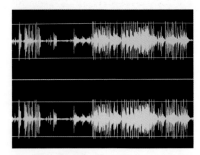

Figure 7.3 *Sound waves captured on audio editing software vary continuously in amplitude and frequency. They are analogue signals.*

Analogue and digital signals

All electronic devices create electrical signals that are either **analogue** or **digital**. An analogue signal has a continuous range of values. A digital signal will only have two possible values, 1 or 0.

A car stereo can handle both analogue and digital signals. Music and speech from a radio station varies continuously in both frequency and volume. The sound produced from the loudspeaker is an analogue wave. For the sound waves to be produced by the loudspeaker in a car stereo system, the electrical signal from the amplifier to the loudspeaker must also be an analogue signal.

Most car stereos also have CD players. The sound on a CD is stored digitally. The signal is a series of electrical pulses, and the height of each pulse is the same. This is a digital signal. It does not vary with a continuous range of values.

Figure 7.4 *A digital signal from a CD.*

A digital signal has only a maximum and minimum value. The maximum value is called 'high' or 'logic 1' and the minimum value is called 'low' or 'logic 0'.

The processing units of car stereos do not only amplify the input signals. They are able to convert the digital signals from digital radio and CDs into analogue signals required by the loudspeaker. Many other electrical systems use both analogue and digital signals.

❓ Did you know ...?

Between 2007 and 2012, all the UK analogue TV signals were switched off and replaced with digital TV signals in what was called the 'digital switchover'. Digital signals can maintain their quality over long distances, and can carry more information. This means the signals can carry more programmes and at higher quality.

Many people still listen to analogue radio, despite digital radios gaining popularity. The digital radio switchover will take place when 50% of all radio-listening is digital, and when the digital radio coverage reaches 90% of the current analogue radio coverage.

Figure 7.5 *The rise of Smart TVs and High Definition was aided by the 'digital switchover'.*

Input devices

Input devices usually change different forms of energy into electrical energy.

The microphone

A microphone is an input device that converts sound energy into electrical energy. Connecting a microphone to an **oscilloscope** can show how the volume and pitch of sounds affect the analogue signal. Louder sounds produce more electrical energy, since the amplitude of the analogue signal is increased.

📖 Word bank

- **Oscilloscope**

An oscilloscope is a device for measuring electrical signals that vary over time. The signals are displayed graphically on a screen.

Figure 7.6 *A microphone and the circuit symbol for a microphone.*

🔵 Activity 7.1

 1. Experiment In this activity you will study the function of a microphone. Using an oscilloscope, you will observe how the amplitude and frequency of an analogue sound signal are linked to volume and pitch.

Read the plan carefully then follow the instructions, making sure you follow all of the safety rules as you would normally be expected to do when carrying out practical work.

(continued)

You can work in small groups but you should record your own observations and complete the answers yourself.

You will need: a microphone, an oscilloscope, connecting leads.

Method

    (a) Connect the microphone to the oscilloscope inputs, as shown.

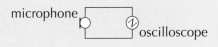

    (b) Whistle into the microphone and adjust the dials of the oscilloscope until you can see a wavelike trace on the screen.

    (c) Adjust the pitch of the whistle by whistling at a higher then a lower note. What happens to the trace on the screen?

    (d) Adjust the volume of the whistle by making it louder then quieter. What happens to the trace on the screen?

  2. What is the energy change in a microphone?

  3. A karaoke system uses a microphone. State whether the microphone is the input, process or output for this system.

> ## 🔍 Hint
> You may require the support of a teacher to complete step (b). Every oscilloscope has different dials in different places, and it takes practice to use them confidently.

> ## Make the link
> The pitch and volume of sound are explored in Chapter 10.

The thermistor

A thermistor is an input device. It does not directly convert energy from one form to another. A thermistor is a type of resistor. The resistance of a thermistor varies with temperature. When the temperature increases, the resistance of the thermistor decreases.

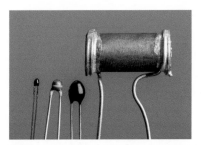
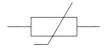

Figure 7.7 *Various thermistors and the circuit symbol for a thermistor.*

Activity 7.2

1. Experiment In this activity you will explore how the resistance of a thermistor varies with temperature.

Read the plan carefully then follow the instructions, making sure you follow all of the safety rules as you would normally be expected to do when carrying out practical work.

You can work in small groups but you should record your own observations and complete the answers yourself.

You will need: a thermistor, an ohmmeter, connecting leads.

Method

(a) Connect a thermistor to an ohmmeter in a circuit as shown.

(b) Record the resistance of the thermistor at room temperature in a table.

Temperature	Resistance (Ω)
Room temperature	
Body temperature	

(c) Warm the thermistor by holding it between your finger and thumb. Record the resistance at body temperature in the table.

2. What happens to the resistance of a thermistor when the temperature increases?

3. Can you think of two electrical systems that could use a thermistor as an input device to respond to changes in temperature?

> ### ★ Memory Aid
> The rule for thermistors can be remembered easily using the acronym TURD. This stands for Temperature Up Resistance Down.

The light-dependent resistor (LDR)

A light-dependent resistor (LDR) is an input device. Like the thermistor, it does not directly convert energy from one form to another. The resistance of an LDR depends on the amount of light incident on it. As the light energy incident on it increases, the resistance decreases.

Figure 7.8 *A light-dependent resistor and its circuit symbol.*

Activity 7.3

1. Experiment In this activity you will explore how the resistance of an LDR varies with light level.

Read the plan carefully then follow the instructions, making sure you follow all of the safety rules as you would normally be expected to do when carrying out practical work.

(continued)

You can work in small groups but you should record your own observations and complete the answers yourself.

You will need: an LDR, an ohmmeter, connecting leads.

Method

(a) Connect an LDR to an ohmmeter as shown.

(b) Under normal room lighting, record the resistance of the LDR in a table like the one below.

Light level	Resistance (Ω)
Light	
Dark	

(c) Now cover the LDR completely, and record the resistance of the LDR.

(d) What happens to the resistance of the LDR as the light incident on it increases?

2. Can you think of two electrical systems that could use an LDR as an input device to respond to changes in light level?

> ### ★ Memory Aid
> The rule for LDRs can be remembered easily using the acronym LURD. This stands for Light Up Resistance Down.

Figure 7.9 *Solar cells and the circuit symbol for a solar cell.*

The solar cell

A solar cell is an input device that converts light energy into electrical energy. The greater the light energy incident on a solar cell, the greater the voltage produced by the cell.

GO! Activity 7.4

1. Experiment In this activity you will investigate the function of a solar cell.

Read the plan carefully then follow the instructions, making sure you follow all of the safety rules as you would normally be expected to do when carrying out practical work.

You can work in small groups but you should record your own observations and complete the answers yourself.

You will need: a solar cell, a voltmeter, connecting leads.

Method

   (a) Connect a solar cell to a voltmeter as shown.

   (b) Record the voltage produced by the cell in a darkened room, or when the solar cell is covered up, in a table like the one below.

Light level	Voltage (V)
Dark	
Light	

   (c) Now illuminate the solar cell by turning on the lights or exposing the cell to sunlight. Record the voltage produced by the solar cell in the light in the table.

   (d) What happens to the voltage produced by the solar cell as the light level increases?

   2. What is the energy change that takes place in a solar cell?

   3. Can you think of two electrical systems that make use of this energy change?

Output devices

Output devices change electrical energy into another form of energy.

The loudspeaker

A loudspeaker is an output device that converts electrical energy into sound energy. Loudspeakers are analogue devices, as the output energy and frequency can vary. Both the output volume (loudness) and frequency can be controlled.

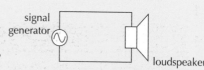

Figure 7.10   *A pair of loudspeakers and the circuit symbol for a loudspeaker.*

GO! Activity 7.5

1. Experiment In this activity you will explore the output of a loudspeaker.

Read the plan carefully then follow the instructions, making sure you follow all of the safety rules as you would normally be expected to do when carrying out practical work.

You can work in small groups but you should record your own observations and complete the answers yourself.

You will need: a signal generator, a loudspeaker, connecting leads.

Method

   (a) Connect a signal generator to a loudspeaker as shown.

   (b) Turn on the signal generator and vary the amplitude dial. What happens to the volume of the loudspeaker?

   (c) Now adjust the frequency dial on the signal generator. What happens to the sound produced by the loudspeaker?

   2. What is the energy change that takes place in a loudspeaker?

   3. A karaoke system uses a loudspeaker. State whether the loudspeaker is the input, the process or the output of the electronic system.

N3	L3	N4	L4

Figure 7.11 *A buzzer and its circuit symbol.*

Figure 7.12 *An electric motor with its internals exposed, alongside the circuit symbol for a motor.*

The buzzer

A buzzer is a device that converts electrical energy into sound energy. It differs from a loudspeaker in that the sound it produces cannot be varied.

Buzzers are used as the audible output in door entry systems, alarm clocks and many other electronic systems.

The motor

An electric motor is an output device that converts electrical energy into kinetic energy. Motors are analogue devices because the speed of the motor can take a range of values.

GO! Activity 7.6

1. Experiment In this activity you will investigate the electric motor.

Read the plan carefully then follow the instructions, making sure you follow all of the safety rules as you would normally be expected to do when carrying out practical work.

You can work in small groups but you should record your own observations and complete the answers yourself.

You will need: a low voltage variable power supply, a motor, connecting leads.

Method

(a) Connect a low voltage variable power supply to a motor as shown.

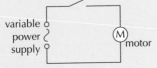

⚠️ Do not turn the voltage supply above the maximum stated voltage of the motor. Doing so can cause damage to the motor and be potentially dangerous.

(b) With the switch closed, gradually increase the voltage of the power supply from 0V to 3V. What happens to the speed of the motor?

2. What energy change takes place in the motor?

3. How many devices can you think of in your home that have electric motors? Make a list of as many as you can.

N3 | L3 | **N4** | L4

The relay

A relay is an output device that converts electrical energy to kinetic energy. A relay is a switch that is controlled by an electromagnet. A relay is a digital device because it can only be open or closed.

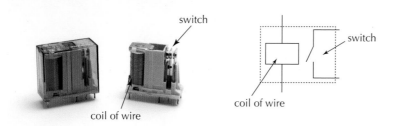

Figure 7.13 *A relay switch and its circuit symbol, both labelled.*

When charges flow through the relay coil, it becomes magnetic. The magnetic field causes the switch to close. Figure 7.14 shows how a relay can be used in an electronic system. The circuit uses a switch connected to a 6V battery to turn a lamp in a 230V circuit on and off.

When the switch is closed, there is a current in the coil of wire. This creates a magnetic field which causes the relay switch to close, lighting the lamp in the high voltage circuit. If a person needs to take the switch apart, they are not exposed to the high voltage circuit. Using a relay switch means high voltage circuits can be operated safely.

The solenoid

A solenoid is an output device that converts electrical energy into kinetic energy. It is another device that uses electromagnetism to operate. A solenoid has a metal core surrounded by a coil of wire.

When a charge flows through the coil of wire, it generates a magnetic field. This causes the metal core to move. It is a digital device as the movement is in one direction and the core is either pushed out or pulled in.

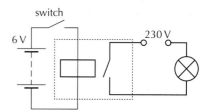

Figure 7.14 *A relay switch connected between a low-voltage and a high-voltage circuit.*

? Did you know ...?

The first relays were used in the nineteenth century in long-distance telegraph circuits. They repeated the electric signal coming from one circuit to another. This way, signals on telegraph wires could travel long distances.

Figure 7.15 *A Morse code telegraph. Relays allowed the signal from these machines to travel long distances.*

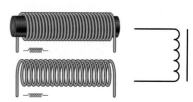

Figure 7.16 *A solenoid and its circuit symbol.*

↻ Keep up to date!

Cars have used solenoids in door locking for many years. Solenoids are also used for medical and industrial purposes.

Search online to find out about how solenoids are used in hospitals and in industrial applications.

Figure 7.17 *A car lock uses a solenoid.*

Figure 7.18 *Different coloured LEDs and the LED circuit symbol.*

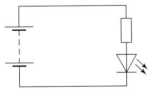

Figure 7.19 *LEDs only work if connected the correct way around in a circuit.*

The light-emitting diode (LED)

Light-emitting diodes (LEDs) are output devices that convert electrical energy into light. LEDs in electrical systems are often used as indicators that an appliance is on. They are used instead of lamps as they only require a small current to operate, and do not give off much heat.

LEDs do not require a high current to operate, and they are usually connected in series with a resistor to limit the current in them. An LED will only allow current to flow in one direction through it. The triangle part of the circuit symbol must point toward the negative end of the battery for the LED to light. Figure 7.19 shows how an LED is connected to a battery.

LEDs are usually digital devices as they are either on or off.

☀ Physics in action: The rise of LEDs

Light-emitting diodes are replacing traditional light bulbs for a variety of applications.

Seven-segment displays are used in many electronic systems for displaying numbers. Each segment consists of 7 LEDs that can be turned on or off to make the numbers 0–9.

LEDs are used for lighting in the home, replacing conventional filament light bulbs and energy-saving fluorescent lamps. They use less electrical energy than filament lamps, largely because they do not produce much heat.

Figure 7.20 *A seven-segment display in a digital clock.*

Figure 7.21 *LED lamps are used in domestic lighting.*

Figure 7.22 *Lightweight LED helmet lamps improve visibility.*

LED lights are replacing conventional lighting in many forms of transport. Bicycles no longer require big heavy batteries to power inefficient lamps.

Many modern cars have daytime running lights that turn on automatically when the car is running. These lights improve visibility during daytime driving, increasing road safety. Clusters of LEDs are fitted in many new cars with this feature to reduce energy usage during the day.

Figure 7.23 *Daylight running lights on modern cars use LEDs.*

Traffic lights also use LEDs instead of lamps. Since LEDs are more reliable, traffic lights fail less frequently.

Figure 7.24 *LED traffic lights.*

GO! Activity 7.7

☺ 1. What are the three parts of an electronic system?

2. Explain the difference between an analogue and a digital device.

3. Name an input device that could detect changes in the temperature of a room.

4. Name an output device that changes electrical energy into kinetic energy.

5. The following table of information is mixed up. Redraw the table, linking each application to the correct device and function.

Application in an electronic system	Device	Function
Output of a radio	Thermistor	To display the cooking time remaining
Input of an automatic lamp	Motor	To detect a change in temperature
Input of a heating controller	Seven-segment display	To change the sound to electrical energy
Output of a fan	Microphone	To produce sound
Output of a home electric safe	LDR	To move a locking bolt
Input of a voice recorder	Loudspeaker	To detect a change in light level
Output of an oven timer	Solenoid	To allow the fan to move

📖 Word bank

- **Transistor**

A transistor is an electronic switching device. A transistor will switch on at a certain voltage. This allows them to be switched on by other electrical components.

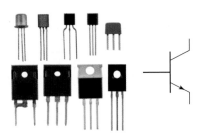

Figure 7.26 *Transistors of various kinds and the transistor circuit symbol.*

❓ Did you know ...?

The founder of Intel electronics, Gordon Moore, is well known for the observation known as Moore's Law. This observation states that transistors will continually be made smaller, so that the number of transistors per square inch on a computer chip will double each year. This trend has resulted in computers that can fit in our pockets (smartphones) being millions of times more powerful than all the computers used to send people to the moon in 1969!

💥 Make the link

Look back to page 111 to remind yourself of how an LDR operates.

Switching devices

Learning intentions

In this section you will:

- learn how a transistor works as a switching device in electronic circuits to solve problems
- describe the properties of logic gates
- state that logic gates and transistors are switching devices.

Transistors

Electronic systems are designed to process the signals from various inputs in order to make decisions. For example, outdoor street lamps are designed to switch on automatically when it gets dark. An LDR is used as the input device, and the lamp is the output device. A device called a **transistor** is used to process the input and switch on the lamp.

Figure 7.25 *The block diagram for an automatic night light.*

A transistor is represented by the circuit symbol seen in Figure 7.26.

A simplified circuit diagram for a street lamp is shown in Figure 7.27. The left-hand side of the circuit contains the input device, the LDR. When it gets dark, the voltage across the LDR goes up to a high enough voltage to allow the transistor to turn on. This causes the lamp to turn on.

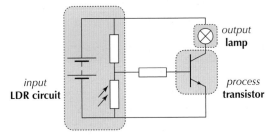

Figure 7.27 *A simplified circuit drawing for a street lamp. The input, process and output stages are highlighted.*

Thermistors can be used instead of LDRs in input circuits, to sense changes in temperature instead of light level. Such a circuit could be used to automatically turn on a heater if it gets too cold.

GO! Activity 7.8

1. Experiment In this activity you will investigate the basic operation of an electronic system that uses a transistor as a switching device.

You can work in a small group but you should record your own observations and complete the answers yourself.

Remember! You must follow the normal safety procedures when carrying out practical work.

You will need: a DC power supply, a transistor switch board, a 10 kΩ resistor, an LDR and a thermistor, ice, tissue paper.

Method

(a) Set up the transistor board so that the LDR is connected at the bottom and the 10 kΩ resistor at the top. Connect the transistor board to the DC power supply and switch it on.

 Help from a teacher may be required to set up the boards correctly.

(b) Observe what happens to the LED (the output device) when the LDR is covered (simulating night time).

(c) Replace the LDR with a thermistor. With the thermistor in the bottom position, it will need to be cooled down to activate the transistor. Hold a block of ice wrapped in tissue paper against the thermistor to cool it down and observe what happens to the LED.

2. Can you think of any applications for a device that turns on when it gets cold? Describe how the electronic system would work.

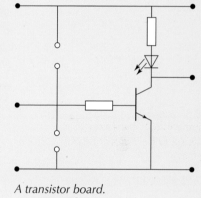

A transistor board.

Logic gates

Many electronic circuits are able to process the signals from multiple inputs and make decisions. For example, a home security system requires two inputs to sound an alarm:

*   the security system needs to be activated (switched on)
*   a sensor needs to detect motion.

If the system is deactivated, the alarm will not sound even if a sensor detects motion. If, on the other hand, the system is activated, the alarm will not sound unless a sensor detects motion. Both inputs need to provide a signal and then the system can process it. Electronic systems process signals in this way using digital switches known as **logic gates**.

Since logic gates are digital devices, they require inputs to be digital also. The inputs can be **high** or **low**. A high is also

Figure 7.28 *Home security systems use electronic logic to keep a home secure.*

📖 Word bank

*   **Logic gate**
    A logic gate enables an electronic system to make decisions based on one or more input.

called 'logic 1' and a low can be called 'logic 0'. The output from a logic gate is also digital.

Although logic gates require digital inputs, they can operate with analogue input devices, such as thermistors, microphones and LDRs. In such cases, the input from the analogue input device is converted to a digital input by the logic gate itself.

There are three basic types of logic gate, called the **NOT** gate (or inverter), the **AND** gate and the **OR** gate.

The function of a logic gate can be displayed in a table known as a **truth table.**

Word bank

• **Truth table**

Truth tables help us to understand the behaviour of a logic gate. They show how the input (or inputs) of a logic gate relate to its output(s).

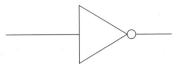

Figure 7.29   *The NOT gate.*

The NOT gate

A NOT gate is the simplest type of logic gate. It inverts the input signal and provides the opposite output. If a NOT gate receives a logic 0 input, it will output a logic 1 (i.e. the output is **not** the input). This is why the NOT gate is also known as an **inverter**.

The truth table for a NOT gate is shown below.

Table 7.2   *Truth table for a NOT gate.*

Input	Output
0	1
1	0

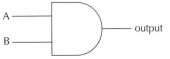

Figure 7.30   *The AND gate.*

The AND gate

An AND gate has two inputs. The output of the AND gate is logic 1 only when both inputs A **and** B are at logic 1. If either input is logic 0, the output is logic 0.

The truth table for an AND gate is shown below.

Table 7.3   *Truth table for an AND gate.*

Input A	Input B	Output
0	0	0
0	1	0
1	0	0
1	1	1

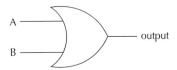

Figure 7.31   *The OR gate.*

The OR gate

An OR gate has two inputs. The output of the OR gate is logic 1 when either input A **or** input B is at logic 1. The output is 0 only when both inputs are logic 0.

The truth table for an OR gate is shown below.

Table 7.4 *Truth table for an OR gate.*

Input A	Input B	Output
0	0	0
0	1	1
1	0	1
1	1	1

Using sensors and logic gates in electronic circuits

National 4
Curriculum level 4
Forces, electricity and waves: Electricity: SCN 4-09c

Learning intentions

In this section you will:

- use your knowledge of electronic components and switching devices to contribute to the development of practical solutions to problems

- investigate the replacement of series/parallel switching in car electrics by logic gates.

By combining logic gates, circuits can be designed to perform useful functions. The examples below show how logic gates can be used.

Example 7.1

A young boy is prone to sleep-walking. His concerned mother designs a circuit to sound a buzzer when he gets up in the night. The circuit uses a light sensor to detect light from the bedroom light and a pressure switch under the rug which switches on when the boy walks on the rug.

The logic gate uses an OR gate.

The buzzer will sound when either the light sensor detects the light on in the son's bedroom, or the switch under the rug is turned on when the son steps on it:

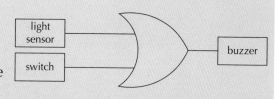

- the light sensor gives a logic 1 when the light is switched on

- the switch outputs a logic 1 when it is switched on.

Draw the truth table that shows all the possible outputs.

Light sensor	Switch under rug	Buzzer
dark (0)	off (0)	off (0)
dark (0)	on (1)	on (1)
light (1)	off (0)	on (1)
light (1)	on (1)	on (1)

Example 7.2

A simple home security system is designed to detect intruders with a light sensor. The intruder will block the light and darken the sensor. The system should only operate when it is first turned on by the home owner.

The system uses a NOT gate and an AND gate.

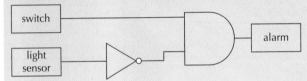

When the system is turned on, the switch will provide a logic 1.

When an intruder is present, the light sensor will be darkened and output a logic 0.

The NOT gate connected to it will invert this to output a logic 1.

When the NOT gate gives a logic 1 output, both inputs to the AND gate are logic 1.

The AND gate will output a logic 1 and the alarm will sound.

Draw the truth table that shows all the possible outputs.

Light sensor	NOT gate for light sensor	Switch	Alarm
(dark) 0	1	(off) 0	(off) 0
(dark) 0	1	(on) 1	(on) 1
(light) 1	0	(off) 0	(off) 0
(light) 1	0	(on) 1	(off) 0

Example 7.3

A gardener wants to protect his young plants in the greenhouse if it gets too cold at night. A system is designed to turn on an electric heater (via a relay) when that happens.

The system uses two NOT gates and an AND gate.

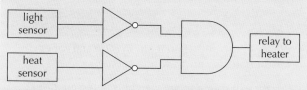

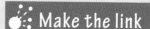

Make the link

Refer to page 115 for a description of how a relay works.

The heater will turn on only when it is dark and cold.

When it is dark the light sensor gives a logic 0.

When it is cold the heat sensor gives a logic 0.

These inputs to the NOT gate will give a logic 1 output.

The AND gate then provides a logic 1 output which will allow the relay to activate. This turns on the heater.

Draw the truth table that shows all the possible outputs.

Light sensor	NOT gate for light sensor	Heat sensor	NOT gate for heat sensor	Relay to heater
(dark) 0	1	(cold) 0	1	(on) 1
(dark) 0	1	(warm) 1	0	(off) 0
(light) 1	0	(cold) 0	1	(off) 0
(light) 1	0	(warm) 1	0	(off) 0

GO! Activity 7.9

😊😊 Many schools have logic gate decision modules like the one shown here.

By connecting the logic board to a battery supply, test the electronic systems described in the three example circuits described above. Why not design an electronic system of your own and ask your partner to explain how it works?

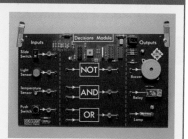

A logic gate investigation board.

Physics in action: Logic gates in car electrics

In Chapter 6, systems for heating and lighting in car electrics were explained using series and parallel circuits. Logic gates are used to provide the function for many useful features of modern cars. Some examples are given in the table below.

Feature	Logic gates used	Description of electronic system
Automatic windscreen wipers	AND gate	Will operate only when ignition is on **and** wiper sensor detects rain on windscreen.
Seatbelt warning sound	AND and NOT gates	Buzzer will sound when weight is detected in seat **and** seatbelt is **not** fastened.
Doors closed detection	AND, OR and NOT gates	Buzzer will sound when engine is on **and** when driver **or** one of the passengers doors is **not** closed.

GO! Activity 7.10

1. Transistors and logic gates are examples of what type of device?

2. The output signal from a temperature sensor can be either logic 0 or logic 1. State which output would be produced if

    (a) the temperature was low?   (b) the temperature was high?

3. A light sensor detects that the light level is bright, and produces a logic 1 output. A NOT gate is connected to the light sensor. What is the output from the NOT gate?

4. The logic gates in an electronic system are connected as shown.

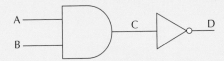

    (a) What is the name of the logic gate connected to inputs A and B?

    (b) What is the name of the logic gate connected to output D?

    (c) Copy and complete the truth table below to show the possible logic levels at positions C and D.

Input A	Input B	C	D
0	0		
0	1		
1	0		
1	1		

N3   L3   **N4**   L4

5. Dusk-till-dawn lamps automatically turn on when it gets dark and turn off at daybreak. Design a system that will switch on a light automatically when it gets dark. There must also be a manual switch.

(Use this layout to help you.)

6. Many fire alarm systems activate if it gets too hot or if smoke is detected. Some systems use a motor to pump water to sprayers attached to the ceiling. Design a system that will turn on a motor when heat from a fire is detected, or when smoke blocks the light to a light sensor.

(Use this layout to help you.)

7. Many cars have warning lights to indicate when the external temperature drops to 0 °C, to alert drivers that there may be ice on the roads.

(a) Design an electronic system that uses a logic gate to alert the driver when the temperature drops to 0 °C, but only when the ignition is on.

(b) (i) State an input device that would be suitable to detect the temperature drop for this system.

(ii) State an output device that could be used as the warning indicator.

Learning checklist

After reading this chapter and completing the activities, I can:

N3 L3 **N4 L4**

- explain the function and application of standard electrical and electronic components. **Activity 7.1, Activity 7.2, Activity 7.3, Activity 7.4, Activity 7.5, Activity 7.6, Activity 7.7 Q3–Q5** ○ ○ ○

- identify analogue and digital input and output devices. **Activity 7.1, Activity 7.2, Activity 7.3, Activity 7.4, Activity 7.5, Activity 7.6, Activity 7.7 Q3–Q5** ○ ○ ○

- state that an electronic system has three main parts: input, process and output. **Activity 7.7 Q1** ○ ○ ○

- explain how a transistor works as a switching device and demonstrate how it is used in electronic circuits to solve problems. **Activity 7.8** ○ ○ ○

N3 L3 **N4 L4**

- explain the use of AND/OR/NOT logic gates in electronic circuits. **Activity 7.10 Q2–Q4**   ○ ○ ○

- state that transistors and logic gates are examples of switching devices. **Activity 7.10 Q1**   ○ ○ ○

- select and use electrical components as input and output devices in practical electronic circuits, providing solutions to real-life situations. **Activity 7.9, Activity 7.10 Q5, Q6**   ○ ○ ○

- describe an example of how logic gates are used in car electrics to perform a certain function. **Activity 7.10 Q7**   ○ ○ ○

8 Gas laws and the kinetic model

This chapter includes coverage of:

N3 Energy transfer • N4 Gas laws and the kinetic model
• Planet Earth SCN 3-04a • Planet Earth SCN 3-05a •
Planet Earth SCN 4-05a • Forces, electricity and waves
SCN 4-08b

You should already know:

- that objects can sink or float in water depending on their properties
- that water can change state from solid to liquid to gas
- that there are different types of energy that can be transferred.

Models of matter

Learning intentions

In this section you will:

- learn about the properties of the three states of matter
- explain how a substance changes state using models of matter
- state that temperature is a measure of how hot or cold something is and is measured in °C
- state that heat is a measure of the amount of energy transferred between a hot and a cold object and is measured in joules
- investigate some ways that changes of state occur in nature
- explain what density is and how it relates to the amount of matter in a certain volume
- learn how to use the relationship density $= \dfrac{\text{mass}}{\text{volume}}$
- investigate how floating and sinking relates to density.

Curriculum level 3

Planet Earth: Processes of the planet SCN 3-04a;
Planet Earth: Processes of the planet SCN 3-05a

Curriculum level 4

Forces, electricity and waves: Forces SCN 4-08b

All matter is made up of particles. The way particles behave depends on the state of matter. The three most common states of matter on Earth are solid, liquid and gas. Table 8.1 shows how the particles in these states are arranged.

Table 8.1 *Particles in a solid, liquid and gas.*

Solid	Liquid	Gas
Particles are in fixed positions. They do not change their positions but vibrate on the spot. Solids are difficult to break apart, due to their strong chemical bonds, and hold their own shape.	Particles are quite close together, but can move around each other. Liquids are quite easy to break apart and do not hold their shape; they take the shape of the container they are in.	Particles have no fixed position in a gas. They move very quickly with lots of space between them. Gas molecules are spaced 10 times further apart than the molecules in liquids and solids. They will spread out to fill the container they are in.

★ You need to know

Temperature and heat are not the same thing. Temperature is a measure of how hot or cold something is. It is usually measured in degrees Celsius (°C). Heat is a measure of the amount of energy transferred between a hot object and a cold object. Heat is measured in joules (J).

When a substance is heated, the energy of the particles increases, so they move faster. The temperature of the substance will also be higher.

All moving particles have energy. The amount a particle moves depends on its energy. When matter is heated, the particles gain more energy and move faster.

Solids, liquids and gases can change state by being heated or cooled. Heating a solid increases the energy of the particles in it. Eventually, the energy of the vibrating particles will be enough to break the chemical bonds holding them together, and the particles will move around, becoming a liquid. Heating them further will increase the particles' energy until they can move apart from each other, becoming a gas.

GO! Activity 8.1

☺
☺☺ **1. Experiment** In this activity you will demonstrate the three states of matter using polystyrene balls to represent particles, and explore how increasing energy changes the state of matter.

Read the plan carefully then follow the instructions, making sure you follow all of the safety rules as you would normally be expected to do when carrying out practical work.

You can work in small groups but you should record your own observations and complete the answers yourself.

You will need: a large number of polystyrene balls and a large tray.

Method

(a) Hold the tray in both hands and gently vibrate it so the polystyrene balls wobble. This represents the energy the particles have in a solid.

(b) Do the particles change position relative to each other when in this state?

(c) Start to move the tray from side to side, allowing the particles to move around the tray.

(d) What state of matter are the particles in now? How does their energy compare with the particles in the solid state?

(e) Now shake the tray both horizontally and vertically until the polystyrene balls move rapidly and bounce off the surface of the tray. The balls now represent the particles of a gas.

(f) How did the energy needed to change the matter into a gas compare with the energy used to change the matter from solid to liquid?

☀ Physics in action: Changes in state all around us

Changes in state occur all around us in nature. They can often be seen easily in water. On cold winter mornings, frost forms on the ground when water vapour (a gas) in the air turns directly into solid ice in a process known as **deposition**. As the Sun warms the ground, the frost **melts**, turning to water. On milder mornings, the water vapour **condenses** to liquid water to form dew on the grass. With continued heating from the Sun, the liquid water will **evaporate**, turning into a gas again.

The **water cycle** is an important natural cycle of changes of states of matter. The Sun's energy **evaporates** water from the sea, changing liquid water to water vapour in the Earth's atmosphere. Water vapour also enters the atmosphere through the process of **transpiration**. Transpiration occurs when plants transfer water from their roots to the underside of leaves where it changes to vapour. Water vapour in the atmosphere cools and **condenses** to form visible clouds. When the water droplets in the clouds get bigger and heavier, **precipitation** takes place, and the water falls as rain, snow or sleet. Snow **melts** on the mountains,

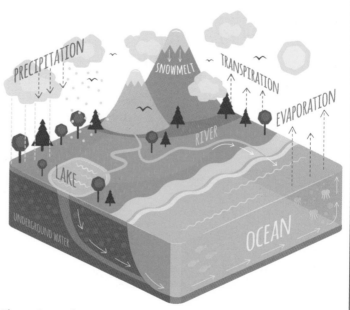

Figure 8.1 *The water cycle.*

(continued)

N3	L3	N4	L4

collecting in rivers along with rain water, and so the water is returned to the sea. The water cycle is an essential process in the maintenance of ecosystems on the planet.

The total mass of water on Earth is relatively constant, but the distribution across rivers, lakes, seas and ice changes with the seasons. Global warming and increasing temperatures at the North and South Poles over the past decade has led to the shrinking of ice sheets and glaciers in Greenland and Antarctica as the ice melts. This is causing a slow but steady increase in sea levels around the world.

Figure 8.2 *Melting icebergs in Greenland caused by global warming.*

Word bank

- **Volume**

Volume is a measure of the amount of space that an object takes up.

Density

Density is a measure of the amount of matter in a certain **volume**.

Density is defined by this equation:

$$\text{density} = \frac{\text{mass}}{\text{volume}}$$

This can be written as

$$d = \frac{m}{v}$$

The units of density are kg/m³ or g/cm³.

Table 8.2 *Symbols and units for the density relationship.*

Name	Symbol	Unit	Unit symbol
density	d	kilograms per cubic metre grams per cubic centimetre	kg/m³ g/cm³
mass	m	kilograms grams	kg g
volume	v	cubic metre cubic centimetre	m³ cm³

The model of particles in matter can explain what density is. A dense material will have more particles in a given volume than a less dense material. For example, polystyrene has a density of 0.024 g/cm³ and lead has a density of around 11.34 g/cm³. The particles in lead are much closer together than those in polystyrene.

Example 8.1

A block of wood has mass of 74 g. The block measures 10 cm × 5 cm × 2 cm.

Calculate the density of the wood.

mass = 74 g

length = 10 cm

breadth = 5 cm

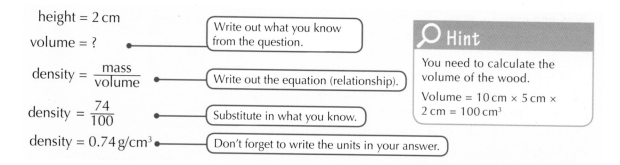

height = 2 cm

volume = ?

Write out what you know from the question.

$$density = \frac{mass}{volume}$$

Write out the equation (relationship).

$$density = \frac{74}{100}$$

Substitute in what you know.

$$density = 0.74 \, g/cm^3$$

Don't forget to write the units in your answer.

Hint

You need to calculate the volume of the wood.

Volume = 10 cm × 5 cm × 2 cm = 100 cm³

GO! Activity 8.2

1. Experiment In this activity you will investigate how the density in a given volume can vary as the size and number of particles is changed.

Read the plan carefully then follow the instructions, making sure you follow all of the safety rules as you would normally be expected to do when carrying out practical work.

You can work in small groups but you should record your own observations and complete the answers yourself.

You will need: a 250 ml beaker, 50 marbles, 100 g of table salt and 100 ml of water.

Method

(a) Fill the beaker with the marbles. Is there room to fit more matter into the same space?

(b) Add the salt into the beaker. What happens to the marbles? What happens to the space between the marbles? What happens to the total mass and density in the beaker?

(c) Finally, pour the water into the beaker. What happens to the marbles and the salt? What happens to the total mass and density in the beaker?

Solids are typically denser than liquids and gases because the particles are more closely packed together. However, some liquids are denser than some solids because of how the particles are arranged in the solid.

The density of an object determines whether it floats or sinks in a liquid. If an object is more dense than a liquid, the object will sink in that liquid. If an object is less dense than the liquid it is in, it will float.

? Did you know ...?

Ice is an unusual material. Most substances are denser in solid form than in liquid form. Usually when liquids cool to form solids, the liquid particles move closer together, forming strong bonds. This means there are more particles in the same volume, making it denser. Ice is different. It is less dense than water, even though it is a solid. For this reason, water pipes are prone to bursting if the temperature drops below 0 °C. The water expands as it freezes, putting pressure on the pipes.

Figure 8.4 *Ice floats in water because it is less dense.*

GO! Activity 8.3

☻ Figure 8.3 shows a column of liquids and solids with different densities. The least dense liquids and solids are at the top, and the most dense liquids and solids are at the bottom.

1. Rank the liquids in order of density, from the least dense to the most dense.

2. Which object is the most dense?

3. Cork is the least dense and floats at the top. Why do you think it has such a low density?

4. Try to create your own density column using the following liquids and objects:

   Golden syrup, water, milk, vegetable oil, an ice cube, a ping pong ball, popcorn kernels.

   Draw a diagram of your density column showing where the objects float.

5. Why do you think ice floats on water even though it is a solid?

Figure 8.3 *A density column with liquids and solids of differing densities.*

GO! Activity 8.4

☻ 1. Name the three states of matter.

2. Describe two differences between solids and liquids.

3. How does the energy of the particles of a liquid compare with those of a gas?

4. Explain the difference between heat and temperature.

5. Give an example from the natural world where changes in the state of matter occur regularly and explain what changes of state occur.

6. An ice cube of mass 15 g is taken from the freezer and heated in a beaker. It quickly melts into a liquid and is eventually boiled off completely. The table shown here contains the densities of water in its three states of matter.

State of matter	Density (g/cm³)
Solid (ice)	0.92
Liquid (water)	1.00
Gas (water vapour)	0.0006

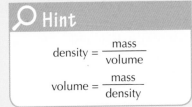

🔍 **Hint**

$$\text{density} = \frac{\text{mass}}{\text{volume}}$$

$$\text{volume} = \frac{\text{mass}}{\text{density}}$$

(a) Use the table of densities to calculate:

   (i) the volume of the ice cube

   (ii) the volume of the water when the ice cube melts

   (iii) the volume of the water vapour.

(b) Comment on how the volume of water in solid, liquid and gas forms compare to each other.

7. The mass of coke in a can of Coca Cola is 374 g. Its volume is 335 cm³.
    (a) Calculate the density of the coke.
    (b) Diet Coke contains much less sugar than Coca Cola. What does this suggest about the density of Diet Coke compared to normal Coca Cola?

8. The density of a cube of iron is 7.8 g/cm³. The mass of the cube is 210.6 g.
    (a) Calculate the volume of the cube of iron.
    (b) Determine the length of one of the sides of the cube.

Heat transfer

Learning intentions

In this section you will:
- learn that heat travels through solids by conduction
- state that metals conduct heat at different rates
- give examples of poor conductors of heat
- learn that heat travels through liquids and gases by convection
- explain what is meant by a convection current
- learn that heat travels in a vacuum by radiation
- state that black objects absorb and radiate heat radiation well, and shiny surfaces reflect radiation well
- investigate examples of heat transfer by conduction, convection and radiation
- learn how buildings can be made energy-efficient by reducing their thermal conductivity.

National 3

Curriculum level 3

Planet Earth: Processes of the planet SCN 3-05a; Planet Earth: Energy sources and sustainability SCN 3-04a

Heat is the transfer of energy from hot objects to cold objects. The way energy is transferred depends on whether it moves through a solid, liquid, gas or a vacuum.

Energy can be transferred by:

- conduction
- convection
- radiation.

Conduction

Heat travels through solids by **conduction**.

Thermal conduction can be explained using a particle model. When a solid is heated, the particles gain kinetic energy and vibrate more. The particles that are heated knock into their neighbours and transfer heat energy to them, causing those particles to vibrate. In this way the heat energy passes along the solid.

📖 Word bank

• **Conduction**

Heat transfer in solids happens by conduction. Heated particles vibrate and the vibrations pass along the particles through the material. Conduction of heat is also called thermal conduction.

Heat from the flame causes the particles to vibrate faster

Vibrations are passed along the conductor

Figure 8.5 *How conduction takes place in solids.*

The ability of a solid to conduct heat is called its **thermal conductivity**.

Some solids conduct heat well, and are good thermal conductors. Metals are examples of good thermal conductors. Generally, if a metal is a good conductor of electricity, it will be a good conductor of heat also.

Other materials are poor conductors of heat, and are good thermal insulators. Wood and plastic are good insulators of heat. Materials that do not conduct electricity are also poor heat conductors. Materials that trap air, such as wool and polystyrene, are also poor heat conductors.

When a good thermal conductor is touched, it feels cold. This is because heat transfers easily from the hand to the conductor. Walking barefoot on ceramic or stone floor tiles feels colder than walking on a carpet. The tiles conduct the heat away from your foot, whereas the carpet is a good insulator because it traps air, which stops conduction taking place.

🔵 Activity 8.5

😊 😊😊 **1. Experiment** In this activity, you will investigate how different metals conduct heat. The activity uses a thermal conductivity bar, which uses temperature indicators on strips of different metals. The different metals are usually steel, aluminium, brass and copper.

Read the plan carefully then follow the instructions, making sure you follow all of the safety rules as you would normally be expected to do when carrying out practical work.

You can work in small groups but you should record your own observations and complete the answers yourself.

You will need: a beaker, a thermometer, a kettle and a thermal conductivity bar.

Method

(a) Bring the kettle to the boil and half-fill the beaker. Place the thermometer in the beaker. Gradually add cold water from the tap until the water in the beaker cools down to 50 °C.

⚠️ Be careful when using hot water.

(b) Lower the thermal conductivity bar into the water.

(c) Observe the heat travelling along each metal bar by looking at the thermal indicators on each bar.

(d) Do all metals conduct heat at the same rate?

(e) Which metal bar conducted heat the quickest? Which conducted heat the slowest?

(f) Each bar is identical in dimensions, but they are different metals. What do you think affects the conductivity of each metal?

Thermal conductivity bars have heat indicators on them.

Activity 8.6

1. State which of these materials are good thermal conductors and which are good thermal insulators.

   copper   **plastic**   **rubber**   **steel**

   wood   **aluminium**   **wool**   **air**

2. Why are saucepans made of steel but have handles made of plastic?

3. Birds fluff up their feathers in the winter time. Explain why this helps to keep them warm.

4. Using your understanding of particles, explain how heat is conducted through a solid.

5. Why does a wooden table feel warm to the touch, when the metal leg of the table will feel cold?

Convection

Heat energy does not travel through liquids and gases by conduction, but can travel through them by **convection**.

Liquids and gases are both fluids. When a fluid is heated, the particles gain kinetic energy and move around faster. This causes them to spread out and the heated fluid becomes less dense. The less dense fluid rises up above the cooler fluid around it. The cooler fluid moves down to take its place, and begins to heat up. This movement of the fluid is called a **convection current**.

Convection cannot take place in solids because the particles of a solid are in fixed positions. In solids, the particles vibrate more when heated but don't change position. In liquids, the particles don't have fixed positions and so they can move around when heated.

Figure 8.6 and Figure 8.7 show convection currents in water. Both show the warm water rising as it becomes less dense, then being pushed to the side by the water rising below it. The images show how the heat is transferred around the water until it is all heated.

Word bank

• **Convection**

Convection describes how heat is transferred through a fluid. A heated fluid moves away from the heat source, carrying the heat energy with it.

Figure 8.6 *Convection current visible by use of a purple dye.*

Hint

Hot air rises, not heat. The movement of the hot air transfers the heat.

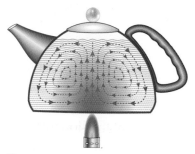

Figure 8.7 *Convection currents in water in a kettle.*

Physics in action: Convection and the weather

Land and sea breezes

Coastal sea breezes are small-scale winds caused by convection currents. On hot sunny days in summer, the sun heats up the land more quickly than the sea. The land warms the air above it, causing it to rise. The air over the sea is cooler and flows in to replace the warm air. On hot days, beaches are often cooler than inland areas due to these sea breezes.

At night, the opposite happens. The land cools quickly, so the sea remains warmer than the land. The sea warms the air above it, causing it to rise. Cooler air from the land is drawn out to sea to replace the warm air. This causes a convection current in the opposite direction.

CONVECTION

Figure 8.8 *Sea and land breezes.*

Large-scale winds

Wind is caused by the uneven heating of the Earth's surface by the Sun. The poles are cooler than the equator because the Sun's rays enter at a more slanted angle there than near the equator. This causes the air near the equator to be warmer and to rise, and the air at the poles to become cooler and sink. The colder polar air moves southward to replace the warmer rising tropical air, setting up a global air circulation.

Ocean currents

The North Atlantic drift is a powerful warm flow of water across the North Atlantic. It is an extension of the Gulf Stream that flows up the east side of North America. It brings warm water from western Florida, and is partly responsible for the mild climate in Scotland compared to the rest of Europe in winter. The North Atlantic Drift is driven by the density difference between the warm water in the south and the lower density, cooler waters in the north. This large flow of water from America to Europe is an example of a large-scale convection current.

Figure 8.9 *The uneven heating of the Earth causes convection currents which create the wind.*

Figure 8.10 *The North Atlantic drift brings the warm waters of the Gulf Stream to Scotland.*

Activity 8.7

1. State what is meant by convection.
2. When a liquid is heated, the liquid rises.
   (a) Explain why this happens using the words **kinetic energy** and **density**.
   (b) What is the name given to the movement of a heated liquid?
3. Explain why anyone standing near a bonfire might feel a draught near their feet.
4. Freezer display units are often left open in supermarkets without having a lid on them. Why does the freezer unit not warm up?
5. Draw a labelled diagram to show how cool sea breezes are formed during a hot summer's day.

Radiation

The space between the Sun and the Earth is a **vacuum**. There are no particles in a vacuum, so heat cannot travel from the Sun to the Earth by conduction or convection. Heat from the Sun travels to the Earth as **radiation**.

Infrared radiation does not need a substance to travel through, unlike heat transferred by conduction or convection. All hot objects emit infrared radiation. When the infrared radiation hits an object, it is absorbed or reflected depending on the colour and the reflectivity of the object. Black objects absorb and emit radiation best, and shiny objects reflect radiation.

Thermal-imaging cameras detect infrared radiation emitted from hot objects and produce images called thermograms. The amount of radiation emitted from a hot object depends on the temperature of the object, so thermograms are used to identify hot and cold areas on objects. Thermal-imaging cameras have many applications including weather forecasting, astronomy, medical diagnostics and monitoring industrial processes.

Word bank

- **Vacuum**
  A vacuum is a space where there are no particles at all.

- **Radiation**
  Heat radiation is also called infrared radiation. This type of radiation is an invisible form of light, and travels in straight lines from a heat source in all directions.

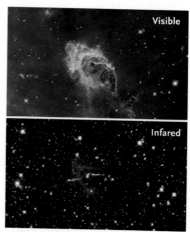

Figure 8.12 *The Carina Nebula Pillar. Infrared light can pass through dust and gas, allowing infrared cameras to detect the stars hidden inside.*

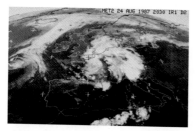

Figure 8.11 *Weather satellites use infrared images to get information about cloud heights and types, and measure land and water temperatures.*

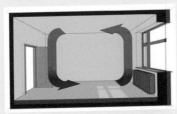

Figure 8.15 *A thermal image shows heat being lost through the walls and windows of a home.*

Reducing the rate of heat transfer

Houses lose heat continually by conduction, convection and radiation. Losing heat from houses and other buildings is a waste of energy and money, so it is a good idea to try to reduce heat loss as much as possible. Thermal imaging is a useful way of seeing where the heat loss is most severe.

Table 8.3 *Heat loss in the home.*

Conduction	Convection	Radiation
Through the floor, walls, windows and roof.	Cold air entering through doors and windows, and convection currents, will transfer the heated air to the roof.	Through the walls, windows and roof.

Heat loss in buildings can be significantly reduced with proper insulation:

- Laying insulation in the attic reduces the conduction of heat through the ceiling.

- Heat loss through windows can be reduced with double glazing. The air gap between the two sheets of glass stops conduction of heat through the window.

- Fitting curtains and draught excluders around doors and windows reduces cold air entering the house.

- Adding cavity wall insulation reduces conduction through the walls of the house.

- Placing foil backing behind radiators reflects the infrared radiation back into the building and stops it passing through the walls.

⟳ Keep up to date!

Housing initiatives in Scotland aim to create new homes which are very energy-efficient. Providing heating and hot water in a typical family home in Scotland costs around £1 100 a year. The Glasgow House is a housing project in Glasgow. It aims to reduce the heating and hot water costs to £100 a year with highly insulated and air-tight homes. Use an internet search engine to find out how well these projects have worked, and investigate other eco-friendly houses built in Scotland.

Figure 8.17   *The Glasgow House.*

GO! Activity 8.8

☺ **1. Experiment** In this activity you will investigate ways of reducing heat loss in houses.
☺☺ A model house can be used to compare heat loss in the following scenarios:

- single glazing vs double glazing
- no loft insulation vs loft insulation
- hollow walls vs cavity-filled walls.

The model house has a lamp inside it which acts as a heat source.

Read the plan carefully then follow the instructions, making sure you follow all of the safety rules as you would normally be expected to do when carrying out practical work.

You can work in small groups but you should record your own observations and complete the answers yourself.

You will need: a model house, a variable power supply, and two thermometers.

Method

(a) Set up the house for your chosen scenario. For example, if you want to test the effectiveness of wall insulation, place insulation in one wall and no insulation in the other. Connect the lamp in the house to the variable power supply and place thermometers in each room. Place the roof on the house.

(b) Turn on the power supply to the lamp. The heat from the lamp will transfer through the walls to the rooms with the thermometers. Record the temperature on each thermometer every minute for a duration of 20 minutes in a table.

A model house for testing heat loss.

(c) Plot a line graph to show how the temperature in each room varied with time.

(d) How did the temperature rise in the room with insulation compare with the room without insulation? Is it what you expected?

(e) Which of the scenarios used to reduce heat loss listed earlier do you think would be the most effective? Explain your answer.

The lamp inside is used as the heat source.

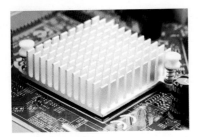

Figure 8.18 *A heatsink on a computer microchip.*

Figure 8.19 *The fan on this heatsink exhausts hot air quickly.*

Increasing the rate of heat transfer

Insulating homes aims to reduce how quickly heat is lost. By contrast, there are some systems which are designed to increase the rate at which heat is transferred.

Computer components such as processors and graphics chips create heat. If they are not cooled, they overheat and malfunction. To prevent this, a heatsink is attached to a processor to draw heat away by conduction. Some heatsinks have fins to increase surface area. This increases the rate at which heat is conducted to the surrounding air. Many heatsinks have fans attached to help draw the hot air away from the processor.

Online computer servers used by large internet companies for storing data need to be kept cool to run efficiently. Air is drawn into the server rooms via a fan and as it flows past the hot servers heat is transferred to the air by conduction. The hot air then rises by convection, collecting at the top of the building where it is then cooled in cooling towers as shown in Figure 8.20. Other data centres are located near water sources. The heat from the servers is transferred to the water, causing it to evaporate.

Figure 8.20 *Cooling towers in a data centre.*

♻ Keep up to date!

Search online to explore what new methods are being used to cool data centres efficiently.

GO! Activity 8.9

☺ 1. Describe how heat transfer by infrared radiation is different to conduction and convection.

2. Explain why a shiny teapot will keep tea hot for longer than a black teapot.

3. Why must heat only travel through space by infrared radiation, and not conduction or convection?

4. Explain why most houses in hot Mediterranean countries are painted white.

5. Describe three ways in which heat can be lost in the home and how the heat loss can be reduced.

6. Use the internet to find out other uses for infrared imaging cameras.

Gas laws and the kinetic model

National 4

Curriculum level 4

Planet Earth: Processes of the planet SCN 4-05a

Learning intentions

In this section you will:

- describe an ideal gas
- learn how to use the relationship $p = \frac{F}{A}$ to determine the pressure exerted by a force on a surface
- learn how the particles in an ideal gas can be explained using the kinetic model
- investigate the effects of varying pressure, volume and temperature on a fixed mass of an ideal gas
- investigate applications of the kinetic model.

The kinetic model of an **ideal gas** is a theory used to describe the behaviour of gases. The kinetic model considers an ideal gas with the following properties:

- the gas is made up of billions of tiny particles (which can be atoms or molecules)

- the particles are moving very fast (for example, air molecules travel around 500 m/s at room temperature)

- the particles are moving in random directions

- the particles are very far apart from one another (relative to the size of the particles)

- the particles can collide with each other and the walls of the container they are in.

There are three main variables to consider in a gas:

- **Pressure**: When the particles of a gas collide with the walls of a container, they exert a force on the walls. This force exerted over the area of the container walls is the **pressure**. The amount of pressure depends on how often and how hard the particles hit the walls. More frequent collisions, or harder collisions, will increase the pressure. Pressure is usually measured in units of newtons per square metre.

- **Volume**: The volume of the gas is a measure of the space it takes up. This usually is the volume of the container holding the gas.

- **Temperature**: The temperature of the gas depends on how much kinetic energy the particles have. They move faster when they have more kinetic energy. When a gas is heated, the particles gain kinetic energy.

📖 Word bank

- **Ideal gas**

An ideal gas is a gas in which no energy is lost in each collision, and in which there are no forces of attraction between each particle.

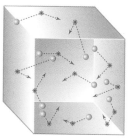

Figure 8.21 *The particles in a gas move about randomly and collide with the walls of their container.*

📖 Word bank

- **Pressure**

Pressure is a measure of the force exerted by an object on a given area. For a gas, the force is exerted by the particles hitting the walls of the container.

Pressure, force and area

When an object exerts a force on another object, the force is spread across the entire surface area that is in contact with the object. For example, a wooden block sitting on a table will exert a force on the table top due to its weight. If the block is then turned on its end, the same force will be spread over a smaller area.

The amount of force exerted on a unit area is the pressure.

Pressure is defined by this equation:

$$\text{pressure} = \frac{\text{force}}{\text{area}}$$

This can be written as

$$P = \frac{F}{A}$$

The units of pressure are N/m^2 or Pa.

> ## ? Did you know ...?
> Air pressure decreases with altitude. As you ascend a high mountain, the amount of air above you decreases. This can result in altitude sickness during climbing expeditions.

> ## 📖 Word bank
> • **Atmospheric pressure**
> This is the pressure exerted by the weight of the atmosphere. At sea level this is around 101 000 pascals.

Table 8.4 *Symbols and units used in the pressure relationship.*

Name	Symbol	Unit	Unit symbol
Pressure	p	Newtons per square metre or pascals	N/m^2 or Pa
Force	F	Newtons	N
Area	A	Square metres	m^2

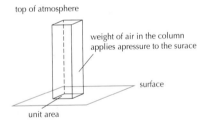

top of atmosphere

weight of air in the column applies a pressure to the surace

surface

unit area

Figure 8.22 *The air in the atmosphere exerts a pressure on the surface of the Earth.*

For a gas, the pressure is the force that the gas exerts on the walls of the container it is in. The gases of the air exert a force called **atmospheric pressure**. The pressure that the atmosphere exerts on the Earth's surface is 101 000 pascals. This is the force that is applied on 1 m^2 on the surface of the Earth by the weight of the air above the surface.

Example 8.2
Determine the weight of the air in a column directly above your hand, when held out flat. Assume your hand has dimensions approximately 10 cm x 20 cm.

Force = weight = ?

area = 0.10 x 0.20 = 0.02 m^2

pressure = 101 000 Pa ⟵ (Write out what you know from the question.)

$$p = \frac{F}{A}$$ ⟵ (Write out the equation (relationship).)

$$101\,000 = \frac{F}{0.02}$$ ⟵ (Substitute in what you know and solve to find the force.)

Weight = 2020 N ⟵ (Don't forget to write the units in your answer.)

Activity 8.10

 1. Experiment In this activity, you will investigate how the force applied to two syringes connected together is affected by the diameter of the syringes.

Read the plan carefully then follow the instructions, making sure to follow all of the safety rules as you would normally be expected to do when carrying out practical work.

You can work in small groups but you should record your own observations and complete the answers yourself.

You will need: two syringes of different diameters, connected together with a rubber tube with water in the tube.

Method

(a) With a partner, squeeze each plunger in. Which one is the easiest to squeeze?

(b) Explain your observation using your understanding of pressure, force and area.

Two syringes connected together with tubing. The syringes are filled with water.

Activity 8.11

1. Experiment Ask your teacher to show you a demonstration of the kinetic model of a gas. Virtual demonstrations can be found online, although the Kinetic Theory Apparatus can also be used.

Method

(a) The dial is turned up, causing the ball bearings to move faster in the tube.

(b) As the speed of the balls increases, what happens to the kinetic energy of the balls?

(c) In a gas, what impact does an increase in kinetic energy have on the temperature of the gas?

(d) What effect does increasing the speed of the ball bearings have on the polystyrene stopper?

(e) What happens to the volume and the pressure when the balls speed up?

The Kinetic Theory Apparatus demonstrator.

The ball bearings in the Kinetic Theory Apparatus represent particles in a gas. The balls move quickly due to the vibrations of the base plate. Each ball bearing hitting the polystyrene stopper exerts a small force on it. Due to the combined force of each ball hitting the stopper, the force is large enough to push the stopper upward. This force on the surface area of the polystyrene is equivalent to the pressure of the ball bearings. This is similar to how gas particles behave in a gas.

Pressure and volume

The apparatus in Figure 8.23 can be used to investigate how pressure varies with volume, for a fixed mass of gas. The volume of the gas is reduced by pushing the plunger in.

When the volume of the syringe is reduced, the gas pressure increases. When the syringe is pushed in, the gas particles have less room to move, as the volume of the gas has been decreased. The particles of the gas will collide more frequently with the walls of the container. Each collision applies a force on the container walls. More collisions mean more force, so the gas pressure will increase. This is true for a fixed mass of gas at a constant temperature.

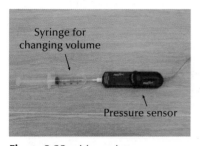

Figure 8.23 *Measuring pressure with a syringe.*

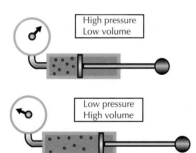

Figure 8.24 *Pressure increases as volume decreases for a fixed mass of gas.*

Activity 8.12

1. Experiment In this activity you will investigate how the pressure varies with volume, for a fixed mass of gas.

Read the plan carefully then follow the instructions, making sure you follow all of the safety rules as you would normally be expected to do when carrying out practical work.

You can work in small groups but you should record your own observations and complete the answers yourself.

You will need: a pressure sensor connected to a computer, a syringe.

Method

(a) Connect the pressure sensor to a computer to display the pressure of the gas.

(b) Push the syringe in to decrease the volume of the gas. What happens to the pressure reading on the computer?

(c) Pull the syringe out to increase the volume of the gas. How does the pressure of the gas change now?

(d) What variables must stay constant in this experiment?

(e) Try to explain how the pressure changes with volume using the kinetic model of gases.

☀ Physics in action: Volume and pressure

The relationship between pressure and volume has useful real-life applications.

In a can of deodorant, the contents are combined with a gas under such high pressure that the gas turns into a liquid. When the nozzle on a deodorant can is pressed, the pressure decreases quickly, causing the gas to expand. The gas forces its way out of the nozzle, carrying the active ingredients of the deodorant with it.

Figure 8.25 *Decreasing pressure in a deodorant can increases the volume of the gas.*

When a syringe is used to draw a solution from a vial, the volume of the syringe increases as the piston is pulled out. This reduces the pressure of the gas in the syringe. The pressure difference between the syringe and the liquid in the vial draws the liquid into the syringe.

Figure 8.26 *Drawing liquid into a syringe.*

Scuba divers need to consider pressure and volume carefully. When a diver descends into the sea, the pressure of the water increases. Due to this high water pressure, the nitrogen in the air that the diver breathes dissolves in the diver's blood. If the diver ascends to the surface too quickly, the rapid pressure drop causes the nitrogen to expand and form bubbles in the body. This can cause a painful condition known as the bends, which is potentially life-threatening.

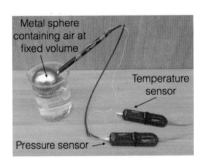

Figure 8.27 *Divers must ascend slowly to avoid a condition known as the bends.*

Pressure and temperature

The apparatus in Figure 8.28 can be used to investigate how pressure varies with temperature, for a fixed mass of gas. This apparatus allows the volume of the gas in the metal sphere to stay constant.

The metal sphere containing a gas is immersed in water of different temperatures. Heat is conducted from the water to the gas, causing the temperature of the gas to rise. The pressure and temperature can be measured by connecting the pressure and temperature sensors to a computer.

As the temperature of the gas increases, the pressure also increases. For an ideal gas, as temperature increases the kinetic energy of the particles also increases. This means they move faster, colliding with the walls of the container more often and with more force. This greater force results in an

Figure 8.28 *Measuring how pressure changes with temperature.*

| N3 | L3 | N4 | L4 |

increase in gas pressure. This is true for a fixed mass of gas at constant volume.

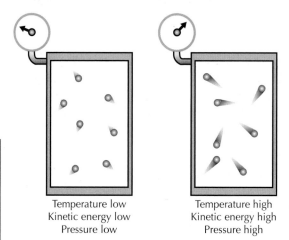

Temperature low
Kinetic energy low
Pressure low

Temperature high
Kinetic energy high
Pressure high

Figure 8.29 *Pressure increases as temperature increases for a fixed mass of gas.*

? Did you know ...?

The air pressure in the tyres of a vehicle increases during a journey. This is because the temperature of the air inside the tyres increases.

GO! Activity 8.13

1. Experiment In this activity you will investigate how pressure varies with temperature, for a fixed mass of gas.

Read the plan carefully then follow the instructions, making sure you follow all of the safety rules as you would normally be expected to do when carrying out practical work.

You can work in small groups but you should record your own observations and complete the answers yourself.

You will need: a pressure sensor, a temperature sensor, beaker of cold water, Bunsen burner, metal sphere filled with gas.

Method

(a) Connect the pressure sensor to a computer to display the pressure of the gas. Prepare a large beaker of cold water and place it over a Bunsen burner.

⚠ Take care to observe normal safety procedures when using Bunsen flames.

(b) Place the metal gas sphere in the beaker of water at room temperature. Light the Bunsen flame and heat the water. What happens to the displayed value of the pressure in the metal sphere?

(c) What variables must stay constant in this experiment?

(d) Try to explain how the pressure changes with temperature using the kinetic model of gases.

| N3 | L3 | **N4** | L4 |

Temperature and volume

The apparatus in Figure 8.30 can be used to investigate how the volume of an ideal gas varies with temperature, for a fixed mass of gas at constant pressure.

Air is trapped in the capillary tube by a bead of mercury at constant atmospheric pressure. When the air is placed in hot water, the temperature of the gas rises. The air in the tube is free to expand and the volume of the air column can be measured using the scale.

As the temperature increases, the volume of the gas also increases when the pressure is constant. For an ideal gas, as the particles are heated they gain kinetic energy and move faster. This results in them striking the walls of the container with more force, and more frequently. In the case where the gas is trapped in a capillary tube by a bead of mercury, the particles strike the mercury. This pushes the mercury upward, increasing the volume. The volume increases until the pressure of the trapped air in the tube returns to its original value.

If the gas were in a container with a movable piston, the piston would be forced up, increasing the volume of the container until the pressure returned to its constant value.

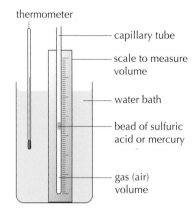

Figure 8.30 *Measuring volume change in a gas.*

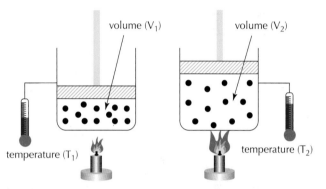

Figure 8.31 *Volume increases as temperature increases for a fixed mass of gas.*

☀ Physics in action: Fixing ping pong balls

A dent in a ping pong ball can be easily fixed. If the dented ball is placed in a pan of water and heated, the dent will quickly pop out. Increasing the temperature of the air in the ball causes the air in the ball to expand, pushing out the dent and restoring the original shape of the ball.

Figure 8.32 *Fixing a dent in a ping pong ball by heating it.*

⊙ Activity 8.14

1. Experiment In this activity you will investigate how the volume varies with temperature, for a fixed mass of gas.

Read the plan carefully then follow the instructions, making sure you follow all of the safety rules as you would normally be expected to do when carrying out practical work.

(continued)

You can work in small groups but you should record your own observations and complete the answers yourself.

You will need: capillary tube, Bunsen burner, beaker of water, bead of mercury.

Method

(a) You could use the apparatus in Figure 8.30 to do this investigation. Place the capillary tube in a beaker of cold water over a Bunsen burner.

⚠ Take care to observe normal safety procedures when using Bunsen flames.

(b) Light the Bunsen flame and heat the water. What happens to the bead of mercury in the capillary tube? What does this mean about the volume of the gas as it is heated?

(c) What variables must stay constant in this experiment?

(d) Try to explain how the volume changes with temperature using the kinetic model of gases.

GO! Activity 8.15

☺ 1. Explain what the term 'ideal gas' means.

2. A pressure gauge used for measuring pressure at different depths in the sea has a cylinder containing a gas. At 50 m below sea level, the pressure of the gas in the cylinder is 850 000 Pa.

   The piston at one end of the cylinder has an area of $0.003 m^2$.

   Calculate the force that the gas exerts on the piston.

3. The pressure at the bottom of a lemonade bottle is greater than at the top. As a bubble rises to the surface, explain what happens to the volume of the bubble.

4. Copy and complete the paragraph below using the words provided to explain why the pressure in a gas rises as the temperature of the gas increases.

force	kinetic energy	collide	pressure	faster
particles	increases		frequently	heated

   When the gas is _ _ _ _ _ _ _ , the _ _ _ _ _ _ _ _ _ _ gain _ _ _ _ _ _ _ _ _ _ _ _ _ _ and move about _ _ _ _ _ _ . This causes them to _ _ _ _ _ _ _ with the walls of the container they are in more _ _ _ _ _ _ _ _ _ _ _ and with greater _ _ _ _ _ . Therefore the _ _ _ _ _ _ _ _ of the gas _ _ _ _ _ _ _ _ _ _ .

5. A child celebrates his birthday in January and is given a helium balloon. He takes it outside into the cold and is disappointed to see it deflates slightly.

   (a) Explain using the kinetic model of an ideal gas why this happens.

   (b) What can the child do to bring the balloon back to its original size?

6. Throwing an aerosol can on a fire may cause it to explode. Use your knowledge of the kinetic model of gases to explain why this can happen.

7. Deep-sea fish die when they are brought too near the surface of the water. Explain why this happens.

Learning checklist

After reading this chapter and completing the activities, I can:

N3 L3 N4 L4

- state that the three most common states of matter on Earth are solid, liquid and gas. **Activity 8.1, Activity 8.4 Q1** ◯ ◯ ◯

- state that changes of state require energy transfer. **Activity 8.1, Activity 8.4 Q3** ◯ ◯ ◯

- explain that density is a measure of the amount of matter in a certain volume. **Activity 8.2** ◯ ◯ ◯

- state that the density of an object is determined with the equation density = mass/volume. **Activity 8.4 Q6–Q8** ◯ ◯ ◯

- state that less dense objects float on substances that are more dense. **Activity 8.3** ◯ ◯ ◯

- state how changes of states of matter are involved in nature. **Activity 8.4 Q5** ◯ ◯ ◯

- state the difference between heat and temperature. **Activity 8.4 Q4** ◯ ◯ ◯

- state that heat is transferred through solids by conduction. **Activity 8.5** ◯ ◯ ◯

- state that different metals conduct heat at different rates. **Activity 8.5** ◯ ◯ ◯

- state that materials like wood and plastic are poor conductors of heat. **Activity 8.6 Q1** ◯ ◯ ◯

- state that heat travels through liquids and gases by convection. **Activity 8.7 Q1** ◯ ◯ ◯

- state that movements of heated particles in liquids or gases are known as convection currents. **Activity 8.7 Q2(b)** ◯ ◯ ◯

- state that heat can transfer through a vacuum by radiation. **Activity 8.9 Q1, Q3** ◯ ◯ ◯

- describe examples of heat transfer by conduction, convection and radiation. **Activity 8.6 Q2, Q5 Activity 8.7 Q3–5, Activity 8.9 Q3, Q5** ◯ ◯ ◯

N3 L3 N4 L4

- state that black objects absorb and radiate heat well and that shiny objects reflect infrared radiation. **Activity 8.9 Q2, Q4**   ○ ○ ○

- identify ways in which buildings can be made more energy-efficient by reducing heat loss. **Activity 8.8, Activity 8.9 Q5**   ○ ○ ○

- state what is meant by an ideal gas. **Activity 8.14 Q1**   ○ ○ ○

- use the relationship $p = F/A$ to determine the pressure exerted by a force on a surface. **Activity 8.14 Q2**   ○ ○ ○

- state that as the pressure of a fixed mass of an ideal gas increases, the volume decreases (at constant temperature). **Activity 8.11, Activity 8.14 Q3**   ○ ○ ○

- state that as the temperature of a fixed mass of an ideal gas increases, the pressure increases (at constant volume). **Activity 8.12, Activity 8.14 Q4**   ○ ○ ○

- state that as the temperature of a fixed mass of an ideal gas increases, the volume increases (at constant pressure). **Activity 8.13, Activity 8.14 Q5**   ○ ○ ○

- state that there are applications of the kinetic model for an ideal gas in daily life and nature. **Activity 8.14 Q5–Q7**   ○ ○ ○

Unit 1 practice assessment

National 3 Outcomes

N3 Energy sources

1. What is the difference between renewable and non-renewable energy sources? **2**

2. Name the three fossil fuels. **1**

3. Name three different renewable energy sources. **1**

4. What is the energy change for a wind turbine? **1**

5. Give an advantage and disadvantage for hydroelectric power stations. **2**

6. Why should we try to reduce our use of fossil fuels? **2**

7. Which is the best renewable energy resource to use in your school? Explain your answer. **2**

N3 Electricity

8. The following information has been taken from four power rating plates on different appliances.

Toaster 1100 W	Iron 800 W	TV 120 W	Kettle 2500 W

Which appliances have a power rating less than 1000W? **1**

9. The following information was found on the rating plate of a steam iron:

Model no.23100

2200 W

220–240 V

32 515 A J

State the power rating of the iron and explain what you could change that would affect the amount of energy transferred by the iron. **2**

10. An energy bill shows an initial reading of 35 476 and a final reading of 35 874. If the unit cost of electricity is 8 p, calculate the cost of the energy used. **3**

11. A current is detected in a coil of wire when a magnet is moved through it. Suggest two ways to increase the current in the coil. **2**

12. Two sets of Christmas tree lights are purchased for decorating the tree. The circuits of the sets are different, as shown in the diagram.

    (a) Identify the set of lamps connected in series. **1**

(b) State an advantage of having the lamps connected as they are in set B, as opposed to how they are connected in set A. 1

(c) A lamp in set A breaks when the switch is closed. Explain why all the lamps turn off. 1

13. The thickness of the copper wires used in household wiring can vary depending on the current drawn by the appliances in the electrical circuit. The table below contains information about the diameter of copper wires used for different electrical circuits in the home.

Circuit	Fuse rating (A)	Copper wire diameter (mm)
Lighting circuit	6	0.255
Electric shower circuit	15	
Small ring mains circuit	20	0.613
Electric oven circuit	30	0.807

(a) Estimate the diameter of copper wire that would be required for the electric shower circuit. 1

(b) Suggest a reason why the diameter of the copper wire used is larger for circuits that draw more current. 1

N3 Energy transfer

14. The diagram below shows how heat transfers by conduction in a solid. Use the diagram to explain how conduction in solids works to transfer heat. 2

metal bar

particle

heat

15. One radiator can heat all the air in a room. Explain how convection currents allow this to happen. 2

16. How is heat transferred from the Sun to the Earth? 1

17. For each of the following methods of reducing heat loss in a home, state and explain how the heat loss is reduced.

(a) Loft insulation.   **(b)** Foil-backed radiators.   **(c)** Double glazing. 1

National 4 Outcomes

N4 Generation of electricity

1. What is the full energy change for a thermal power station?   1

2. What is the function of a turbine in a thermal power station?   1

3. Why are thermal power stations not 100% efficient?   1

N4 Electrical power

4. State the definition of power.   1

5. A hairdryer of power rating 1000 W and a convection heater of power rating 25 000 W were both switched on for 15 minutes. Which one transfers the most energy?   3

6. A food processor transfers 12 000 J of energy in 30 seconds. Calculate the power of the appliance.   3

7. A hydroelectric power station produced 60 000 W of power. The input power is 67 000 W. Calculate the efficiency of the power station.   3

8. What can be done to help save electrical energy in our homes? Give three examples.   3

N4 Electromagnetism

9. Sketch the magnetic field pattern around the following arrangements of magnets. Include arrows to show the direction of the magnetic field.

   (a)

N   S

      1

   (b)

N S		N S

      1

   (c)

S N		N S

      1

10. A wire is wrapped around an iron nail and connected to a variable voltage power supply. State two ways in which the strength of the electromagnet could be increased.   2

11. The following circuit is set up using a coil of wire and a cell.

    Sketch the magnetic field pattern around the coil of wire when the switch is closed.   1

12. State an application of an electromagnet and explain how it operates.   2

13. The figure shows a transformer used in a charger for a tablet device. Name the parts labelled A, B and C. **3**

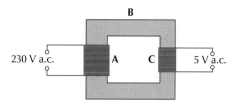

N4 Practical electrical and electronic circuits

14. This circuit contains four resistors.

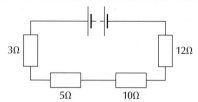

Calculate the total resistance in the circuit. **3**

15. A student is using the circuit below to investigate current and voltage in resistors. The student uses a cell, an ammeter, a voltmeter and two resistors, R_1 and R_2.

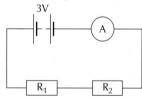

   (a) Copy and complete the diagram to show where the student would connect a voltmeter to measure the voltage across resistor R_1. **1**

   (b) The student finds that the voltage across resistor R_1 is 1.2 V. Calculate the voltage across resistor R_2. **3**

   (c) The student suggests to his teacher that moving the ammeter closer to resistor R_1 will give a more accurate value of the current through this resistor. Explain why the student is incorrect. **2**

   (d) The ammeter reads 0.050 A. Calculate the resistance of resistor R_2. **3**

N4 Gas laws and the kinetic model

16. The kinetic model of a gas describes how the particles in a gas behave. State three things the kinetic model describes the particles doing. **1**

17. When a gas is heated, what effect does this have on the particles of the gas?

18. The statements below relate to the three most common states of matter.

   The particles are very close together.

   The particles move about quickly.

   The particles are far apart and move in all directions.

   The particles do not move about, they just vibrate.

   The particles are quite close together.

   The particles move about.

   Draw a table like the one below, and place the statements under the appropriate heading. In the bottom row, draw a diagram of each state of matter showing how the particles behave.   **3**

SOLID	LIQUID	GAS

19. A road safety website states the following in an article about tyre pressure:

   'Aim to check pressures every couple of weeks, with the tyres cold, using a reliable and accurate tyre pressure gauge.' Source: The AA

   Why is it important to check the tyre pressure when they are cold, and not when they are hot?   **2**

20. During a flight, a passenger notices that the packet of crisps he has purchased before the flight appears to have more air in it than when he bought them. Explain why this has occurred using the words **pressure** and **volume**.   **2**

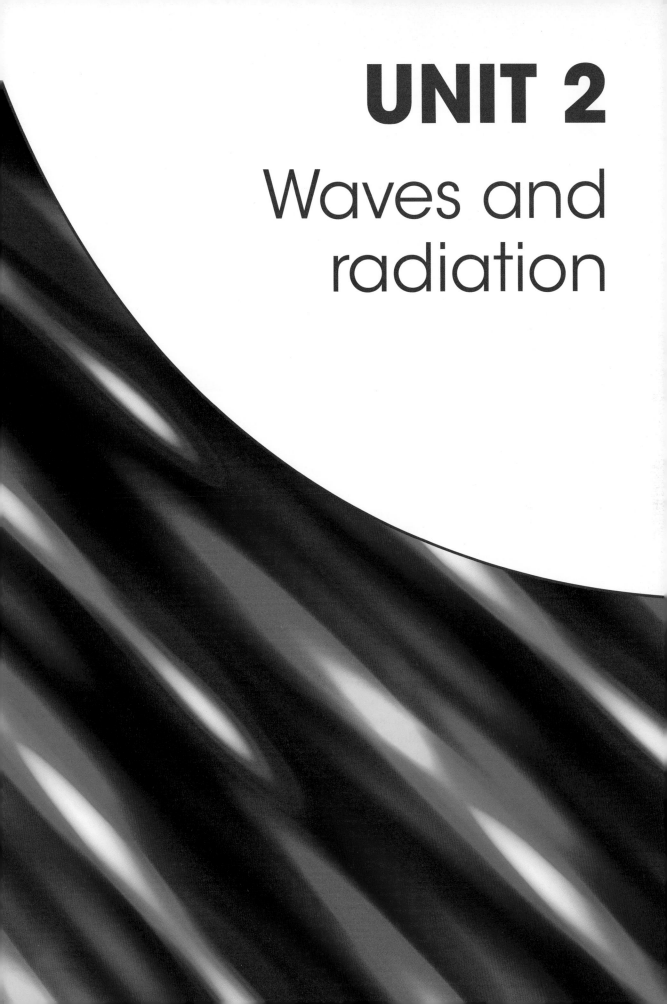

UNIT 2

Waves and radiation

9 Wave characteristics

This chapter includes coverage of:

N3 Wave properties • N4 Wave characteristics

You should already know:
- that sound waves are produced by vibrations
- that sound can travel through different media.

Waves and energy

Learning intentions

In this section you will:
- learn that waves transfer energy
- name and identify correctly features of a wave, such as wavelength, crest, trough and amplitude.

> **Make the link**
>
> Electric, magnetic and gravitational fields are explained more in Chapter 15.

Waves are created by vibrations, generated by objects or fields moving backwards and forwards or up and down. Waves carry energy. Different types of waves transfer different types of energy, such as kinetic energy, sound energy and light energy.

Physics in action: Catching a wave!

Surfers line up just outside the breaking waters and wait for bigger waves. Here, they bob up and down as the smaller waves pass them. As they 'catch' a wave, the surfers gain kinetic energy that has been transferred from the wave.

Figure 9.1 *Surfers waiting to 'catch' a wave.*

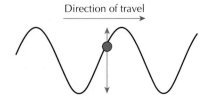

Direction of travel

Figure 9.2 *A float in the water bobs up and down as a wave passes.*

When a boat or a fisherman's float is sitting in the water, it will bob vertically up and down as a wave passes. The wave itself moves horizontally.

Wind and tides create water waves which transfer kinetic energy. This kinetic energy is potentially damaging to ships at sea and to buildings and structures on shore. The waves in severe storms can be dangerous. Storm barriers are built to protect harbours from damage.

The kinetic energy in water waves can also be used to generate electricity.

The height of a wave is proportional to the energy of the wave, so if the energy being transferred by the waves decreases, the waves will be smaller.

Waves are described using special terms. You need to know these terms and use them when you describe waves; these are called **wave characteristics**.

Figure 9.3 *Potentially damaging waves in Aberdeen Harbour.*

> **Make the link**
>
> Using waves to generate electricity is explained in more detail in Chapter 1.

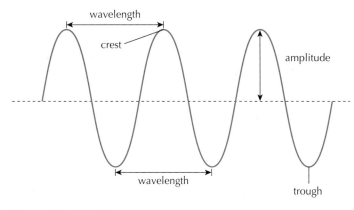

Figure 9.4 *Labelling a wave.*

Table 9.1 *A list of wave characteristics.*

Term	Description
Crest	The highest point of a wave.
	Sometimes the crest of a wave can be referred to as the **peak**.
Trough	The lowest point of a wave.
Amplitude	Amplitude is a measure of the height of a wave. It is measured from the middle of the wave to a crest or a trough and is half the total height of the wave.
	Amplitude is one measure of the energy a wave carries. This bigger the amplitude, the greater the energy.
Wavelength	The distance between two identical points on a wave, such as two successive crests.
Frequency	The number of waves that pass a fixed point per second.

GO! Activity 9.1

Answers to all activity and assessment questions in this Unit are available online at www.leckieandleckie.co.uk/page/Resources

☻ **1.** Identify the parts of the wave labelled A, B, C and D.

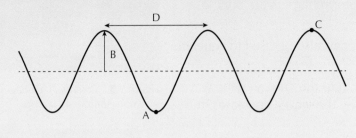

GO! Activity 9.2

☻☻ **1. Experiment** In this activity you will use a sheet of paper and a swinging source of sand or paint to produce and label waves.

Read the plan carefully then follow the instructions, making sure you follow all of the safety rules as you would normally be expected to do when carrying out practical work.

You should work in a small group but record your own observations and complete the answers yourself.

You will need: a paper cup, a clamp and stand, a pin, string, some tape, fine sand and a long sheet of paper. (You can use diluted paint instead of sand.)

Method

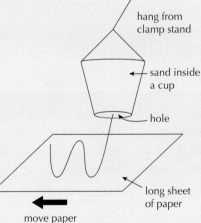

(a) Use a pin to make a small hole in the bottom of a paper cup. Make sure the hole is big enough for the sand to pass through freely. Make two more holes on opposite sides near the rim.

(b) Tie a string through the top holes and hang the cup from a clamp stand.

(c) Cover the bottom hole with a piece of tape. Fill the cup with sand or paint.

(d) Place a long sheet of paper under the cup.

(e) Pull the cup back to the edge of the paper, remove the tape from the hole, and allow the cup to swing.

(f) At the same time, drag the paper at constant speed in a direction at right angles to the direction of the cup's swing.

(g) Trace the shape of the wave on the paper. Label a crest, a trough, the amplitude and the wavelength.

(h) Investigate what happens to the wave pattern when you change the length of the pendulum and the speed at which the paper is moved.

2. How can you obtain a wave pattern that has the same wavelength but different amplitude?

3. How can you obtain a wave pattern that has the same amplitude but a shorter wavelength?

↻ Keep up to date!

Floating weather stations known as buoys (pronounced 'boys') record weather data in the seas and oceans around the world. Buoy K5 is sited about 270 km off the Isle of Lewis in the Outer Hebrides in the North Atlantic. This buoy has recorded wave heights of 19 m.

Search online or use the Met Office marine observations webpage to find out the height of today's waves around Scotland's coast. Share your findings with others and discuss if you think these developments in technology have had an impact on modern life.

Types of waves

National 4

Learning intentions

In this section you will:

- learn that there are two types of waves, longitudinal and transverse
- learn how to recognise longitudinal waves
- learn how to recognise transverse waves.

Many types of energy can be carried as waves. There are two types of wave:

- **longitudinal waves**, such as sound waves and seismic waves (earthquakes)
- **transverse waves**, such as water waves, and light and radio waves.

Both types of waves are formed by vibrations. For example, sound waves are formed by vibrating particles, and transfer sound energy. Light waves and radio waves are vibrating electric and magnetic fields, and transfer electromagnetic energy.

Longitudinal waves

When a wave passes through a material the particles in the material vibrate. If the particles move in the same direction as the wave, the wave is classified as a longitudinal wave.

Sound waves are longitudinal waves. A loudspeaker moves backwards and forwards causing air particles to move together and then away from each other. This creates the sound wave.

📖 Word bank

- **Longitudinal waves**

In a longitudinal wave the particles move in the same direction as the wave.

- **Transverse waves**

In a transverse wave the particles move at right angles to the direction the wave is travelling.

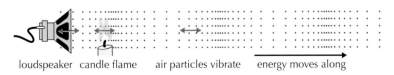

loudspeaker   candle flame   air particles vibrate   energy moves along

Figure 9.5 *Particles in a longitudinal wave. The vibrations can be seen by the flickering flame.*

💥 Make the link

Sound waves are described in more detail in Chapter 10.

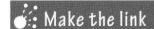

| N3 | L3 | **N4** | L4 |

Transverse waves

In transverse waves the particles move at right angles to the direction the wave is travelling. For example, the particles could move up and down while the wave is transferring energy from left to right.

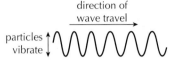

direction of wave travel →

particles vibrate ↕

Figure 9.6 *A transverse wave.*

Water waves are an example of transverse waves. Light waves are transverse waves and belong to a group of waves called **electromagnetic waves**. Electromagnetic waves do not need a material to travel through, but they cause electric and magnetic fields to vibrate.

GO! Activity 9.3

👥 **1. Experiment** In this activity you will use a slinky to examine the two main types of waves and how they transfer energy.

Read the plan carefully then follow the instructions, making sure you follow all of the safety rules as you would normally be expected to do when carrying out practical work.

You can work in a small group but you should record your own observations and complete the answers yourself.

You will need: a slinky spring and a board or flat surface.

Method

  (a) Attach one end of the slinky to the board (or get a friend to hold the end still).

  (b) Hold the slinky at the other end and move it in and out (forward and backward), expanding and contracting the coils. This is a **longitudinal** wave.

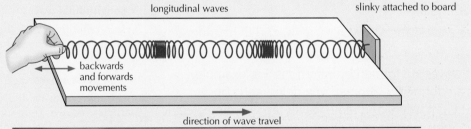

longitudinal waves · slinky attached to board

backwards and forwards movements

direction of wave travel

  (c) Next, move the slinky from side to side. A pulse travels along the wave to the board at the other end, although the individual coils only move sideways. This is a **transverse** wave.

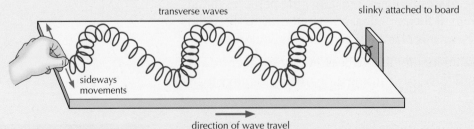

transverse waves · slinky attached to board

sideways movements

direction of wave travel

Report

Write a short description of what you have seen. Include the words transverse and longitudinal.

You can film the motion of the slinky spring to watch the waves in slow motion. Tracker software can pinpoint a coil on the wave and analyse the motion.

Activity 9.4

1. Name three types of energy that are transferred by a wave.

2. The diagram shows a wave. Is it transverse or longitudinal?

3. Explain your answer to Question 2.

4. Give an example of a type of energy that travels as this type of wave.

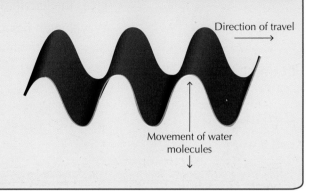

Direction of travel

Movement of water molecules

Wave characteristics

Learning intentions

In this section you will:

- learn how to calculate the amplitude of a transverse wave
- learn how to calculate the wavelength of a transverse wave
- state that frequency is the number of waves per second.

Waves are described in terms of:

- amplitude – this is a measure of the energy transferred by a wave. It is measured on a wave trace from the middle of the wave to a crest or a trough, or half the total height of the wave.

- wavelength – the distance between two identical points on the wave.

- frequency – the number of waves passing a fixed point per second.

- period – the time for a complete wave to pass a fixed point.

Measuring amplitude and wavelength

A wave trace can be used to measure some of the features of a wave.

To measure the **amplitude** of a wave, measure the distance from the mid-point to the top of a crest or the bottom of a trough. This is the same as measuring the total height from the top of the crest to the bottom of the trough and dividing by 2.

Amplitude has units of metres (m).

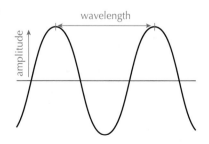

Figure 9.7 *A wave pattern showing the amplitude and the wavelength.*

To measure the **wavelength** of a wave, divide the total distance the wave occupies by the number of complete waves. This gives the length of one wave shape.

Wavelength has units of metres (m).

Example 9.1

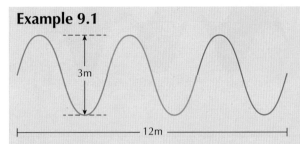

Calculate:

(a) the amplitude of the wave

(b) the wavelength of the wave.

(a) Total height of the wave = 3 m •——(Write out what you know from the question.)

Amplitude = 3 ÷ 2 = 1.5 m •——(The amplitude is the total height divided by 2. Remember to write in the units.)

(b) 3 complete waves in 12 m •——(Write out what you know from the question.)

Wavelength = 12 ÷ 3 = 4 m •——(The wavelength is the distance divided by the number of waves. Remember to write in the units.)

GO! Activity 9.5

☺ 1. Draw a wave shape and label a crest, a trough, the amplitude and a wavelength.

2. The wave in the diagram has a total height of 2.5 m. Calculate the amplitude.

3. The wave in the diagram has a total length of 12 m. Calculate the wavelength.

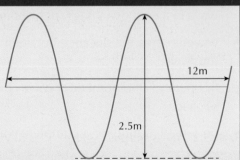

4. If 20 waves occupy a total space of 100 m, calculate the wavelength of one wave.

5. A guitar string is plucked and starts to vibrate. The total length of the string is 0.6 m and there are 2 complete waves along the string. Calculate the wavelength of the waves.

6. A harp string is plucked and starts to vibrate. The total length of the string is 1.2 m and there is half a complete wave along the string. Calculate the wavelength of the waves.

7. Sam and Finn are investigating the waves on a slinky spring. Finn waves the end of a slinky while Sam records the motion using a camera. On the recording there are 12 waves in a distance of 4 metres. Calculate the wavelength.

8. Katie jumps into the swimming pool at the deep end and starts splashing. She creates waves that have a wavelength of 1 m. If there are 50 waves along the whole length of the pool, calculate the length of the swimming pool.

9. Search online to find an online wave simulator. Experiment with the settings that let you change the wavelength and amplitude of waves.

Wave frequency

Learning intentions

In this section you will:

- explain what is meant by the frequency of waves
- use the relationship between frequency, number of waves and time
- use the relationship between wave speed, frequency and wavelength
- use the relationship between distance, speed and time for waves.

Frequency is a measure of how often something happens. This term can be used when describing waves.

Frequency is measured in hertz (Hz). A frequency of 10 Hz means that 10 waves pass a fixed point in 1 second. This is 10 waves per second.

This definition can be written as an equation.

$$\text{frequency} = \frac{\text{number of waves}}{\text{time taken}}$$

This can be written as

$$f = \frac{N}{t}$$

📖 **Word bank**

- **Frequency**
Frequency is the number of waves that pass a fixed point in 1 second.

Table 9.2 *Terms and units used to calculate frequency.*

Name	Unit	Unit symbol
frequency	hertz	Hz
number of waves	no units	no units
time taken	seconds	s

Example 9.2

250 waves are produced in 20 seconds. Calculate the frequency.

number of waves = 250

time taken = 20 s ●───────────[Write out what you know from the question.]

frequency = ?

$$\text{frequency} = \frac{\text{number of waves}}{\text{time taken}}$$ ●───────[Write the equation (relationship).]

$$\text{frequency} = \frac{250}{20}$$ ●───────[Substitute in what you know.]

frequency = 12.5 Hz ●───────[Don't forget to write the units in your answer.]

GO! Activity 9.6

😊 **1.** Explain what is meant by the terms 'wavelength' and 'frequency'.

2. If 560 waves are generated in 20 seconds, calculate the frequency of the wave.

3. What is the frequency of a sound wave from a loudspeaker that vibrates 452 times in 2 seconds?

4. What is the frequency of a wave from a wave generator that produces 5 waves every second?

5. What is the frequency of a piano string that produces 393 waves in 1.5 seconds?

GO! Activity 9.7

😊😊😊 **1. Experiment** This activity uses a shallow water tray to carry out an experiment to observe water waves.

Read the plan carefully then follow the instructions, making sure you follow all of the safety rules as you would normally be expected to do when carrying out practical work.

You should work in a small group to discuss your observations and ideas.

You will need: a rectangular tray, a wooden dowel or pipe and some water.

Method

Making and observing water waves.

(a) Fill a rectangular tray with water to a depth of about 1.5 cm and place on a level surface.

(b) Use a wooden dowel thicker than 2 cm diameter (or something similar) with a length slightly shorter than the width of the tray. Place in the water at one end.

(c) Roll the dowel back and forth repeatedly and watch the waves moving across the water's surface.

(d) Roll the dowel back and forth at a slower rate (lower frequency) and observe what happens to the wavelength.

(e) Roll the dowel back and forth at a faster rate (higher frequency) and observe what happens to the wavelength.

2. What do you think is the relationship between frequency and wavelength?

3. Find out what happens if you use a greater depth of water in the tray.

4. If you have access to a ripple tank, watch the waves that can be produced. Video analysis can be used to learn more about the behaviour of the waves.

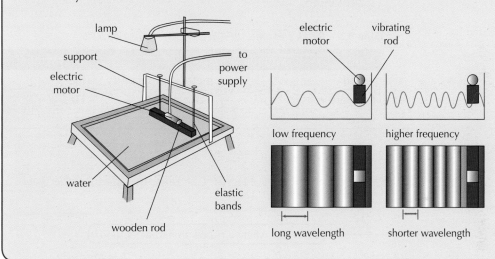

Wave speed, frequency and wavelength

As a water wave passes a fixed point, the water particles move up and down as the wave travels along. The number of wave crests that are seen is related to the **speed** of the wave, the wavelength and the frequency. Many wave crests will be seen when:

- the wave is travelling quickly

- the frequency is high

- the wavelength is short.

There is a relationship between the speed, the frequency and the wavelength of a wave. When the speed of the wave is constant, the frequency goes up if the wavelength goes down, and the frequency goes down when the wavelength goes up.

📖 Word bank

- **Speed**

Speed is a measure of the distance travelled in a given time. Speed has units of metres per second (m/s) or kilometres per hour (km/h).

The relationship between wave speed, frequency and wavelength is called the **wave equation**:

> **speed = frequency × wavelength**

This can be written as

$$v = f\lambda$$

where:

v is used to represent speed

f is used to represent frequency

λ is the Greek letter lambda and is used to represent wavelength.

Table 9.3 *Terms and units used in the wave equation.*

Name	Symbol	Unit	Unit symbol
speed	v	metres per second	m/s
frequency	f	hertz	Hz
wavelength	λ	metres	m

Example 9.3

A wave has a frequency of 500 Hz and a wavelength of 0.5 m. Calculate the speed of the wave.

frequency = 500 Hz

wavelength = 0.5 m ← Write out what you know from the question.

speed = ?

speed = frequency × wavelength ← Write the equation (relationship).

speed = 500 × 0.5 ← Substitute in what you know.

speed = 250 m/s ← Don't forget to write the units in your answer.

Example 9.4

A sound wave has a speed of 340 m/s and a wavelength of 10 m. Calculate the frequency of the wave.

v = 340 m/s

λ = 10 m ← Write out what you know from the question.

f = ?

$v = f\lambda$ ← Write the equation (relationship).

$340 = f \times 10$ ← Substitute in what you know and solve to find the frequency.

$\dfrac{340}{10} = f$

$f = 34$ Hz ← Don't forget to write the units in your answer.

N3 | L3 | **N4** | L4

GO! Activity 9.8

☻ **1.** Calculate the missing values **(a)–(d)** for this table.

Speed (m/s)	Frequency (Hz)	Wavelength (m)
(a)	40	1.2
340	**(b)**	0.2
(c)	5000	2.0
300 000 000	440 000	**(d)**

2. A sound wave has a frequency of 850 Hz and a wavelength of 0.4 m. Calculate the speed of the wave.

3. A wave has a wavelength of 0.56 m and travels at a speed of 24 m/s. Calculate the frequency of the wave.

Calculating wave speed

The speed that a wave travels at depends on the medium it is travelling through. Sound waves travel faster in solids than in liquids or gases, because particles in a solid are packed close together and vibrations are passed along more easily. In gases the particles are spaced further apart, so waves are slower. In deep space, there are no particles, so sound cannot travel.

Different types of waves travel at different speeds. The speed of sound in air is approximately 340 m/s. The speed of light is approximately 300 000 000 m/s.

⚛ Make the link

There is more information on how to calculate and measure speed in Chapter 14.

⚛ Make the link

Find out how to measure the speed of sound in Chapter 10.

✸ Physics in action: Thunder and lightning

During a thunderstorm, a flash of lightning is accompanied by the sound of thunder. The flash of lightning and the sound of thunder are produced at the same time, but the flash of lightning is seen before the thunder is heard, because light travels much faster than sound. You can calculate the distance from the lightning by measuring the time between the lightning flash and the thunder and multiplying the time by the speed of sound.

Figure 9.8 *Lightning strikes.*

The speed at which a wave transfers energy is calculated using this relationship:

$$\text{speed} = \frac{\text{distance}}{\text{time}}$$

This can be written as

$$v = \frac{d}{t}$$

where:
v is used to represent speed
d is used to represent distance
t is used to represent time.

Table 9.4 *Terms and units used in the speed of waves equation.*

Name	Symbol	Unit	Unit symbol
distance	d	metres	m
speed	v	metres per second	m/s
time	t	seconds	s

Example 9.5

Calculate the speed of a water wave that travels a distance of 40 m in a time of 50 seconds.

$d = 40\,\text{m}$
$t = 50\,\text{s}$ — Write out what you know from the question.
$v = ?$

$v = \dfrac{d}{t}$ — Write the equation.

$v = \dfrac{40}{50}$ — Substitute in what you know and solve to find the speed.

$v = 0.8\,\text{m/s}$ — Don't forget to write the units in your answer.

GO! Activity 9.9

☺ **1.** Calculate the missing values **(a)–(d)** for this table.

Speed (m/s)	Distance (m)	Time (s)
260	**(a)**	10
5.5	**(b)**	2
(c)	1000	0.25
27	1350	**(d)**

2. A wave travels 46 m in 10 s. Calculate the speed of the wave.
3. A wave travels 120 metres in 180 seconds and it has a wavelength of 0.25 metres.
   (a) Calculate the speed of the wave.
   (b) Calculate the frequency of the wave.
4. The sound of thunder is heard 4 seconds after seeing a flash of lightning. The thunderstorm is 1380 metres away. Calculate the speed of the sound wave.

Learning checklist

After reading this chapter and completing the activities, I can:

N3 L3 N4 L4

- name the characteristics of a wave such as wavelength, amplitude, crest and trough. **Activity 9.1, Activity 9.2** ○ ○ ○

- state that waves transfer energy. **Activity 9.3** ○ ○ ○

- state that there are two types of waves, longitudinal and transverse. **Activity 9.3, Activity 9.4 Q2** ○ ○ ○

- identify longitudinal waves. **Activity 9.3, Activity 9.4 Q2** ○ ○ ○

- identify transverse waves. **Activity 9.3, Activity 9.4** ○ ○ ○

- calculate the amplitude of waves. **Activity 9.5 Q2** ○ ○ ○

- calculate the wavelength of waves. **Activity 9.5 Q3–Q7** ○ ○ ○

- state that frequency is the number of waves per second. **Activity 9.6 Q1, Activity 9.7** ○ ○ ○

- use the relationship between frequency, number of waves, and time. **Activity 9.6 Q2–Q5** ○ ○ ○

- use the relationship between wave speed, frequency and wavelength. **Activity 9.8 Q1–Q3** ○ ○ ○

- use the relationship between distance, speed and time for waves. **Activity 9.9 Q1–Q4** ○ ○ ○

10 Sound

This chapter includes coverage of:

N3 Sound • N4 Sound • Forces, electricity and waves
SCN 4-11a

You should already know:

- that sound vibrations are carried by waves through solids, liquids and gases.

National 3

National 4

Curriculum level 4

Forces, electricity and waves: Vibrations and waves SCN 4-11a

Sound waves

Learning intention

In this section you will:

- describe sound as a longitudinal wave.

There are sounds all around us. Some are loud, others quiet. They have different tones, beats and pitches. All sounds have one thing in common. They are created by vibrations. When something vibrates (wobbles) it sets the particles around it vibrating. These particles collide with other particles. The kinetic energy of these collisions is carried as a sound wave. Our ears can detect these sound waves and so we hear a noise.

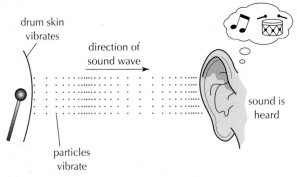

drum skin
vibrates

direction of
sound wave

sound is
heard

particles
vibrate

Figure 10.1 *How sound reaches our ear.*

📖 Word bank

- **Longitudinal waves**

In a longitudinal wave the particles travel in the same direction as the wave is travelling.

Sound waves are **longitudinal waves**. The particles in the wave move in the same direction that the wave is travelling. The vibrating object pushes air particles together and pulls them apart.

Sound waves have two separate properties which affect what we hear.

- **Frequency** is a measure of the pitch of a sound. The frequency is the number of sound waves that pass a fixed point per second.

- **Amplitude** is a measure of the loudness, or volume, of the sound.

When we describe sound, we talk about high-pitched sounds and low-pitched sounds. Sounds will have different frequencies, or pitches. For example, a ship's horn or a note from a bass drum is a low-pitched sound, and whistles and sirens are high-pitched sounds. High-pitched noises have a high frequency. There are more vibrations every second.

When the amplitude of a sound wave increases, the sound gets louder. Increasing the energy in the vibration increases the amplitude of the sound wave. Plucking a string harder, or beating a drum harder, will transfer more energy.

Most everyday sounds are combinations of waves that have different amplitudes and frequencies. In a science lab we can generate pure sound waves that only have a set amplitude or frequency. This can be done by using a signal generator.

📖 Word bank

• **Frequency**
Frequency is the number of waves that pass a fixed point per second. Frequency is measured in hertz (Hz).

🌐 Make the link

See Chapter 9 for more about wave characteristics

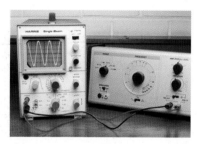

Figure 10.2 *A signal generator and an oscilloscope displaying a wave.*

🔵 Activity 10.1

😊😊 **Experiment** In these experiments you will investigate the vibrations and sounds made by different objects.

Read the plan carefully then follow the instructions, making sure you follow all of the safety rules as you would normally be expected to do when carrying out practical work.

Record your own observations.

😊😊 **Experiment 1**: Sound travelling in solids
You need to do this activity with a classmate.

You will need: two paper cups, about 3 m of string, something to punch a hole in the bottom of the paper cups.
Method
- **(a)** Punch a hole in the bottom of two paper cups.
- **(b)** Attach them together with the piece of string.

(continued)

(c) Each person should hold one paper cup. Walk apart until the string is **tight**. One person should talk into their cup while the other listens in their cup.

(d) Can you hear anything?

(e) How do you think this works? (Hint: think about the particles in the string.)

(f) What happens if the string is not tight?

Experiment 2: Making a wine glass sing

You will need: a wine glass.

Method

(a) Fill the wine glass with some water.

(b) Dip your finger in the glass and run your finger around the top of the glass.

(c) Can you make the wine glass 'sing'?

(d) Why do you think this makes a noise?

Making a wine glass `sing'.

> 🔍 **Hint**
>
> If you do not have access to a wine glass, search online to find a video of someone using this technique to play music. There are some pretty impressive performances!

Experiment 3: Tuning forks

You will need: two tuning forks tuned to the same note, a retort stand, a small ball hanging on a piece of string.

Method

(a) Set up the apparatus as show in the diagram. Make sure the hanging ball is just touching the second tuning fork.

(b) Set the first tuning fork vibrating and watch the ball next to the second fork.

(c) Why do you think the ball bounces? (Hint: think about the particles.)

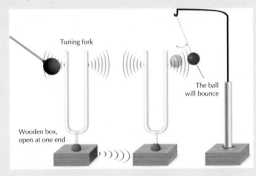

Tuning fork

The ball will bounce

Wooden box, open at one end

Tuning forks vibrating and transferring energy.

> A tuning fork is a two-pronged metal fork that can be set vibrating to produce a sound. Each size fork produces a specific musical note.

Human hearing range

Learning intentions

In this section you will:

- learn about the frequencies that humans and other animals can hear
- measure your own hearing range.

Humans can only hear certain frequencies within a **range of hearing**. The range of hearing defines the lowest and the highest frequencies we can hear. For most people, the lowest frequency they can hear is about 20 Hz, and the highest is about 20 000 Hz (or 20 kHz). The highest frequency we can hear tends to decrease as we get older.

The human range of hearing is different from other animals. Some elephants can hear frequencies as low as 14 Hz. Some breeds of dogs can hear sounds with frequencies up to 45 000 Hz, and some bats can hear sounds with frequencies up to 200 000 Hz.

Word bank

- **Range of hearing**

The range of hearing is the range of frequencies that can be heard.

GO! Activity 10.2

☺ 1. **Experiment** In this activity you will measure your own range of hearing.

Read the plan carefully then follow the instructions, making sure you follow all of the safety rules as you would normally be expected to do when carrying out practical work.

Record your own observations.

You will need: a signal generator, loudspeaker, connecting leads.

A signal generator is an electronic device that generates an electrical signal with a set frequency. The frequency and amplitude of the output of the signal generator can be changed. An oscilloscope is a device for viewing waveforms.

Method

(a) Connect the signal generator to the loudspeaker.

(b) Starting with a frequency of about 100 Hz, gradually decrease the frequency until you can no longer hear a sound. Record this frequency.

(c) Increase the frequency up to about 12 kHz and higher. Keep increasing the frequency until you can no longer hear a sound. Record this frequency.

2. Did everyone in your class have the same range of hearing?

⚠ Do not have the volume too high. It could damage your hearing.

National 3

National 4

Curriculum level 4

Forces, electricity and waves: Vibrations and waves SCN 4-11a

Measuring frequency and amplitude

Learning intentions

In this section you will:

- investigate what happens when the frequency of a sound wave is changed
- investigate what happens when the amplitude of a sound wave is changed
- manipulate and analyse waveforms.

Sound waves can be represented in a diagram as shown in Figure 10.3. The curve, or **waveform**, is a graph of the displacement of the air particles at different distances along the wave. The different features of the sound wave have specific names to describe them:

- the **crest** is the highest point of the wave
- the **trough** is the lowest point of the wave
- the **amplitude** is a measure of the height of the wave, measured from the middle of the wave to a crest or a trough. It is half the total height of the wave.
- the **frequency** is the number of waves that pass a fixed point per second.

Make the link

There is more about wave characteristics in Chapter 9.

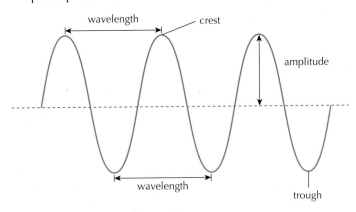

Figure 10.3 *Features of a sound wave.*

Sound waves can be detected using a microphone. The microphone is an input device, so converts the sound energy into electrical energy. This can be shown as a waveform on an **oscilloscope** screen.

Changing the sound alters the waveform. The sound can be changed by:

- changing the frequency
- changing the amplitude.

When the frequency is increased, there are more vibrations every second. This means the waveform has more crests and troughs when viewed on a screen. The pitch of the sound increases. The amplitude of the wave stays the same.

If the frequency is decreased, there are fewer vibrations every second. This means the waveform has fewer crests and troughs when viewed on a screen. The pitch of the sound decreases. The amplitude of the wave stays the same.

> **📖 Word bank**
>
> - **Oscilloscope**
> An oscilloscope is a piece of laboratory equipment used to display waveforms.

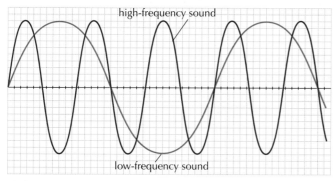

Figure 10.4 *Changing the frequency (pitch).*

When the amplitude is increased, the particles in the wave carry more energy. This means the waveform on the screen is taller. The volume of the sound increases (it gets louder). The frequency (number of waves on the screen) remains the same.

When the amplitude is decreased, the particles in the wave carry less energy. This means the waveform on the screen is less tall. The volume of the sound decreases (it gets softer). The frequency (number of waves on the screen) remains the same.

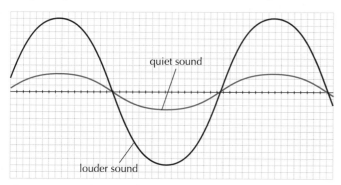

Figure 10.5 *Changing the amplitude.*

Activity 10.3

1. **Experiment** In this activity you will investigate the shape of waveforms using a signal generator and an oscilloscope.

The output of the signal generator is used as the input for a loudspeaker to create a sound. The waveform can be observed on the oscilloscope.

Read the plan carefully then follow the instructions, making sure you follow all of the safety rules as you would normally be expected to do when carrying out practical work.

Record your own observations.

You will need: a signal generator, an oscilloscope, a loudspeaker, connecting leads.

Method
(a) Set up the apparatus as shown in the diagram.

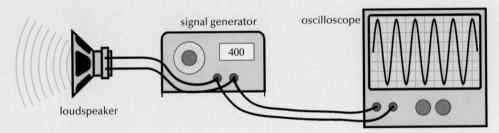

Measuring sound waves.

(b) Listen to the sound and watch the waveform on the screen when the volume of the sound is changed.

(c) Describe what happens to the waveform when the sound is made louder and then quieter.

(d) Listen to the sound and watch the waveform on the screen when the frequency of the sound is changed. Describe what happens to the waveform when the frequency is changed.

(e) If possible, connect your oscilloscope to a piano keyboard or similar instrument and observe the waveforms as you press different keys.

(f) If possible, connect a microphone and observe the waveforms as you sing a song. You could also use a tuning fork to generate a note and compare this to someone singing or whistling the same note.

GO! Activity 10.4

☺ **1.** Look at the waveforms in the diagram.

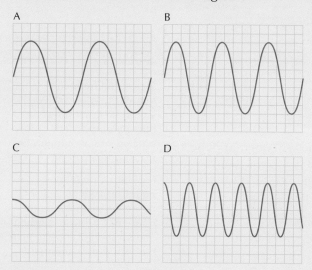

Which of these waveforms show:

(a) a quiet sound?   **(b)** a loud, high-pitch sound?

(c) a low-pitch sound?

2. Explain the term 'frequency'.

3. Describe what happens to the waveform seen on an oscilloscope screen if a sound gets quieter and its pitch is lowered.

? Did you know ...?

In music, the pitch of a note is measured in octaves. If you go up an octave on a musical instrument, the frequency of the note produced doubles. Middle C has a frequency of 256 Hz. The C one octave above that has a frequency of 512 Hz (key shaded green).

Figure 10.6 *Middle C and an octave above on a keyboard.*

? Did you know ...?

When you talk you produce very unique sound patterns. These patterns can be analysed and used to identify your voice. Police detectives can use these patterns to match voice patterns from recordings of known criminals. They also identify patterns from shots of particular guns.

Voice recognition can also be used to control computer software. Some telephone banking systems use voice recognition to check customers making transactions. Search online to find out more about voice recognition software.

Activity 10.5

Experiment Plan an experiment to investigate the frequency of sound. You could:

- use straws cut to different lengths (with a reed cut)
- change the height of water in a bottle
- change the length of a vibrating ruler off the edge of a desk
- use different tuning forks
- use a musical instrument or a sonometer.

Think carefully about the apparatus you will need and the measurements you will have to take. Will you measure frequency **qualitatively** or **quantitatively**? Make sure you follow all of the safety rules as you would normally be expected to do when carrying out practical work. You should work in a small group but record your own observations. Don't forget to evaluate your experiment.

Hint

Measuring frequency **qualitatively** means you would be describing it as being high-pitched, or low-pitched. You would not be measuring anything.

Measuring frequency **quantitatively** means you would be measuring the frequency using a microphone and an oscilloscope to give a numerical measurement.

National 3

National 4

Table 10.1 *The approximate speed of sound in different media.*

Medium	Speed of sound (m/s)
Carbon dioxide	270
Air	340
Water	1500
Human tissue	1500
Muscle	1600
Glycerol	1900
Bone	4100
Aluminium	5200
Steel	5960

Measuring the speed of sound

Learning intentions

In this section you will:

- learn that sound travels at different speeds in different materials
- measure the speed of sound in air
- use the relationship between speed, distance and time to calculate the speed of sound.

Sound waves need particles of matter to transmit energy. The energy is passed to adjacent particles, and the particle collisions cause the sound wave to travel. A sound wave travelling in a single material has a constant speed.

Sound waves travel faster in solids than in liquids or gases, because particles in a solid are packed close together. The material that the waves travel through is called the **medium**. (The plural of medium is **media**.)

The particle theory of matter explains how sound travels in a solid, liquid and gas.

- In a solid the particles are packed closely together. The sound wave travels quickly.

- In liquids the particles are less closely packed than in a solid, but still close enough to collide frequently. The sound wave travels less quickly than in a solid.

- In gases the particles are far apart. The particles do not collide very often and sound travels slower.

- In a vacuum there are no particles. The sound wave cannot travel.

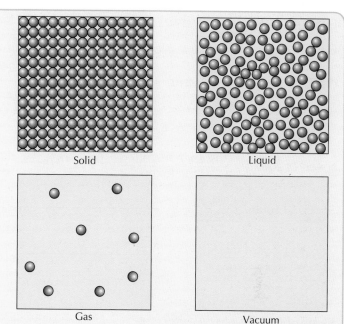

Figure 10.7   The particle theory of matter.

Make the link

In Chapter 8 there is more information about solids, liquids and gases.

The speed of sound in air is around 340 m/s. The time for sound to travel 1 m will be around 3 milliseconds. A time of 3 milliseconds cannot be accurately recorded using a manual stopwatch, so it is much better to use electronic timers for such short times. The speed of sound in liquids and solids is even faster.

Hint

1 millisecond is 1 one-thousandth of a second.

The speed of sound is calculated using this relationship:

$$\text{speed of sound} = \frac{\text{distance travelled}}{\text{time taken}}$$

This can be written as

$$v = \frac{d}{t}$$

Table 10.2   *Terms and units used in the speed of sound equation.*

Name	Symbol	Unit	Unit symbol
distance	d	metres	m
speed	v	metres per second	m/s
time	t	seconds	s

Example 10.1

Calculate the speed of sound for a wave that travels a distance of 68 m in a time of 0.2 seconds.

distance travelled, $d = 68\,m$

time taken, $t = 0.2\,s$ — Write out what you know from the question.

speed, $v = ?$

speed of sound $= \dfrac{\text{distance travelled}}{\text{time taken}}$ or $v = \dfrac{d}{t}$ — Write out the equation (relationship).

speed of sound $= \dfrac{68}{0.2}$ — Substitute in what you know and solve to find the speed.

speed of sound, $v = 340\,m/s$ — Don't forget to write the units in your answer.

Example 10.2

Calculate the time taken for a wave to travel a distance of 50 m if the speed of the wave is 4 m/s.

distance travelled, $d = 50\,m$

speed, $v = 4\,m/s$ — Write out what you know from the question.

time taken, $t = ?$

speed of wave $= \dfrac{\text{distance travelled}}{\text{time taken}}$ — Write out the equation (relationship).

$4 = \dfrac{50}{\text{time taken}}$ — Substitute in what you know and solve to find the speed.

time taken $\times 4 = 50$

time taken, $t = 12.5\,s$ — Don't forget to write the units in your answer.

Example 10.3

Jessica stands in a large cave. She hears an echo 1 second after making a noise. If the speed of sound is 340 m/s, calculate the distance from Jessica to the cave wall.

$v = 340\,m/s$

$t = 1\,s$ — Write out what you know from the question.

$d = ?$

$v = \dfrac{d}{t}$ — Write out the equation (relationship).

$340 = \dfrac{d}{1}$ — Substitute in what you know and solve to find distance.

(continued)

N3 L3 N4 L4

$340 \times 1 = 340\,m$ •————

This is the total distance travelled by the wave, to the wall and back again. The question asks for the distance from Jessica to the wall so the value needs to be halved.

Distance from Jessica to the wall = $340 \div 2$

Distance from Jessica to wall = $170\,m$ •———

Don't forget to write the units in your answer.

❓ Did you know ...?

Some blind people learn to use echoes as a way of locating objects around them. By clicking their tongue and listening to the echoes they can avoid walking into obstacles. This is similar to bats using ultrasound echolocation. Bats can avoid obstacles and find food using the echo in pitch-black darkness.

Figure 10.8 *Bats use echolocation to find food.*

GO! Activity 10.6

👀 **1. Experiment** In this activity you will measure the speed of sound in air. You need two people for this activity. One person will make a loud sound and the second person will record time using a stopwatch.

Read the plan carefully then follow the instructions, making sure you follow all of the safety rules as you would normally be expected to do when carrying out practical work.

Record your own observations.

You will need: two blocks of wood, a stopwatch, a measuring tape or trundle wheel.

Method

   (a) Measure and record the length of a playground or field using a measuring tape or trundle wheel.

   (b) One person stands at the end of the field with two blocks of wood. They clap the two blocks of wood together to make a loud sound.

   (c) The other person stands at the other end of the measuring field with the stopwatch.

   (d) The person with the stopwatch should start the stopwatch when they see the wood-blocks being clapped or the balloon popped, and stop the stopwatch when they hear the noise. Record the time shown on the stopwatch.

   (e) Repeat the experiment a number of times and calculate the average time. This will give more reliable results.

   2. Calculate the speed of sound using the average time in the equation:

   $$\text{speed of sound} = \frac{\text{distance travelled}}{\text{average time taken}}$$

GO! Activity 10.7

😊😊 1. **Experiment** In this activity you will measure the speed of sound in air using electronic devices. The experiment is not subject to human reaction times.

Read the plan carefully then follow the instructions, making sure you follow all of the safety rules as you would normally be expected to do when carrying out practical work.

Record your own observations.
You will need: a hammer, a metal plate, two microphones, a timer with two inputs, connecting leads, metre stick or measuring tape.

Method

(a) Set up the apparatus as shown in the diagram.

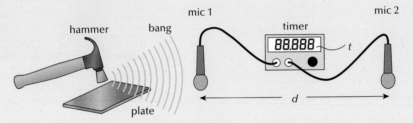

Electronically measuring the speed of sound.

(b) Measure the distance between the two microphones using a metre stick or measuring tape.

(c) Switch on the timer.

(d) Hit the hammer on the metal plate to produce a loud sound near microphone 1.

(e) The timer will start when the sound is detected at microphone 1. It will stop when the sound is detected at microphone 2. Record the time shown on the timer.

2. Calculate the speed of sound using the relationship:

$$\text{speed of sound} = \frac{\text{distance between microphones}}{\text{time taken}}$$

3. How could you use this experimental method to measure the speed of sound in a liquid or a solid? If possible, plan and carry out your designs.

? Did you know ...?

In athletics the sound of a starting pistol is played through loudspeakers. These speakers are the same distance from each runner. This ensures each runner hears the sound of the starting pistol at the same time. When sprints are won and lost by hundredths of a second, it's essential that everyone hears the starting pistol at exactly the same time to avoid any runner getting an unfair advantage.

Activity 10.8

☺ The table shows the speed of sound in different media.

Use the data in the table to answer these questions.

Medium	Speed of sound (m/s)
Carbon dioxide	270
Oxygen	312
Air	340
Water	1500
Lead	1960
Copper	5010
Steel	5960

1. In which material does sound travel the fastest?

2. Does sound travel fastest in solids, liquids or gases?

3. Draw a bar chart of these speeds.

4. Calculate the time taken for a sound wave to travel 25 m through:

   (a) oxygen   (b) water

   (c) lead   (d) copper

5. The sound of a firework is heard 0.6 seconds after the flash is seen. The firework exploded 200 m above the crowd. Calculate the speed of the sound wave.

6. A sound wave travels down a solid steel bar. It takes 0.0025 seconds. Calculate the length of the steel bar.

7. Oliver hears the sound of thunder 6 seconds after seeing a flash of lightning. The thunderstorm is 2070 metres away. Calculate the speed of the sound wave.

8. The equipment shown is used to measure the speed of sound. The two microphones are separated by a distance of 3 m. The time taken for the sound to travel from one microphone to the other is measured to be 0.009 s. Calculate the speed of sound.

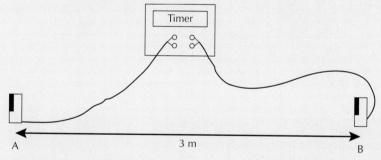

A   3 m   B

9. Tom hears an echo from a wall 2 s after the initial sound. Calculate the distance from Tom to the wall. The speed of sound in air is 340 m/s.

10. A submarine detects an echo from an underwater object 0.3 seconds after it sends out the sound signal. Calculate the distance from the submarine to the object. The speed of sound in water is 1482 metres per second.

Sound level

The loudness of sound is measured with a decibel meter and has units of **decibels** (dB). The sound of complete silence has a value of 0 dB. The human ear can be damaged if sounds are too loud. Table 10.3 shows some typical sound levels.

📖 Word bank

- **Decibels**
Loudness of sound is measured in decibels (dB).

Table 10.3 *Typical sound levels measured in decibels.*

Noise level (dB)	Example
0	Threshold of hearing
20	Quiet bedroom
40	Library
50	Ordinary conversation
60	A typical classroom
70	Passenger car, travelling at 35 miles per hour
80	Traffic at a busy road-side
100	Pneumatic drill
120	Threshold of pain
160	Gun fire

Figure 10.9 *A sound-level meter.*

🔵GO! Activity 10.9

Experiment Use a sound-level meter to measure and record the sound level for different situations and different locations. You could try:
- your classroom
- a busy corridor
- the path outside your school.

⚠ Make sure you follow all of the safety rules as you would normally be expected to do when carrying out practical work.

1. Draw a table to record your observations.
2. Where was the loudest place you measured?

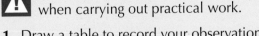

Effects of surfaces

Hard, flat surfaces reflect sound well and produce strong echoes. Soft surface materials that contain lots of air pockets, such as fabrics, foam and sponge, are not good at reflecting sound. They absorb it. The sound waves transfer energy to the air in the pockets so less is reflected.

Some materials can be shaped to reflect sound in different ways. Jagged and curved surfaces reflect sound differently to flat surfaces.

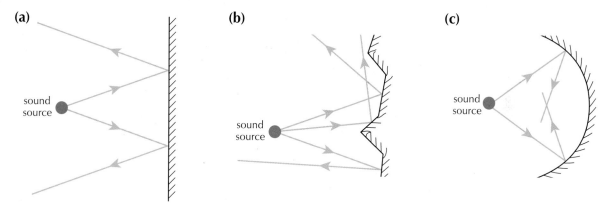

Figure 10.10 *Sound reflecting from different surfaces.* **(a)** *Flat surface* **(b)** *Jagged surface* **(c)** *Curved surface*

GO! Activity 10.10

😊😊 **Experiment** Plan and carry out an experiment to investigate the absorption of sound in different materials. You could use a cell and a buzzer to generate the sound. Could you make a container that is lined with different materials to represent an anechoic chamber? Think carefully about the apparatus you will need and the measurements you will have to take.

Prepare a presentation about your experiment to show to your class.

Make sure you follow all of the safety rules as you would normally be expected to do when carrying out practical work.

You should work in a small group but record your own observations. Don't forget to evaluate your experiment.

? Did you know ...?

Heriot-Watt University and the University of the West of Scotland use **anechoic chambers** to study sound characteristics and perform experiments that need isolating from sound vibrations. The walls of anechoic chambers are designed to absorb all of the sound waves. The room is very soundproof.

Figure 10.11 *The walls of an anechoic chamber.*

Word bank

- **Noise pollution**
Noise pollution is unwanted sounds.

Noise pollution

Sounds that are not wanted are often called **noise**. Noise could be a vacuum cleaner interrupting you when you are doing your homework, the corridor noise during a test or noise from machinery in the work place.

Some noise is impossible to avoid, such as traffic noise on a busy street, or the sound of people talking in a busy shopping centre. This kind of noise doesn't usually cause big problems. At other times, or in other locations, noise can cause problems, by interrupting sleep, causing a noise nuisance or damaging hearing. This is all **noise pollution**.

People can report noise pollution to their local council. Complaints can include:

- loud parties and music
- domestic equipment (e.g. washing machines, vacuum cleaners)
- barking dogs
- street noise, such as arguments and shouting
- burglar alarms and car alarms
- construction activity (e.g. pile driving, hammering, shouting, vehicles).

GO! Activity 10.11

☻ 1. Search online to investigate the noise pollution due to airports. Are flights allowed to take off throughout the day and night? What are the complaints about airport noise? What are the possible solutions to noise pollution in this situation?

Report

Write a short report about the problems of noise pollution from airports and some possible solutions.

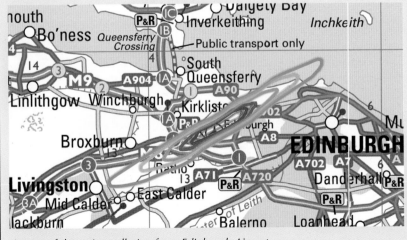

Contour colour	Noise level (dB)
	69
	66
	63
	60
	57

A map of the noise pollution from Edinburgh Airport.

Hearing protection

The ear converts energy from sound waves to electrical impulses that are interpreted by the brain. Very loud sounds produce high-pressure waves which can rupture the eardrum or damage the small bones inside the ear. Hair cells and nerves in the ear can also be damaged by loud noises. The eardrum may heal itself over a long period of time, but hearing damage due to loud noises can be permanent.

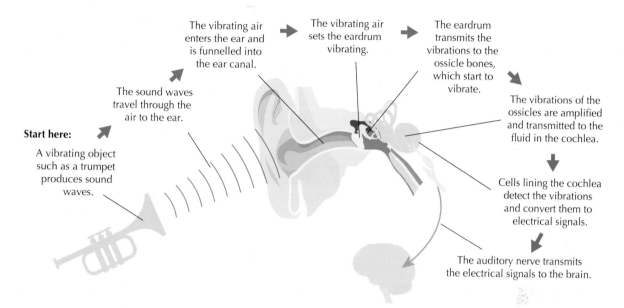

The vibrating air enters the ear and is funnelled into the ear canal.

The vibrating air sets the eardrum vibrating.

The eardrum transmits the vibrations to the ossicle bones, which start to vibrate.

The sound waves travel through the air to the ear.

The vibrations of the ossicles are amplified and transmitted to the fluid in the cochlea.

Start here:

A vibrating object such as a trumpet produces sound waves.

Cells lining the cochlea detect the vibrations and convert them to electrical signals.

The auditory nerve transmits the electrical signals to the brain.

Figure 10.12 *How we hear sounds.*

Excessively loud noise for a prolonged period of time can damage hearing. Noise levels in factories, airports and other places which use heavy machinery have to be monitored to reduce the risk of exposure to excessively loud noise levels.

Employers have to provide ear protection to workers if their workers are exposed to noise levels above 85 dB for a period of time. Examples of hearing protection include:

* wearing ear defenders
* turning down the volume of sound-making devices
* increasing the distance from the source of noise
* reducing the time of exposure to loud sounds
* using soundproofing material.

Flying in a helicopter is a very noisy environment. Pilots and passengers in helicopters will often wear **noise-cancelling headphones**. Noise-cancelling headphones work in this way:

- a microphone picks up the background sounds

- these sounds are converted to an electrical signal, which is then inverted (turned upside down)

- the inverted signal is fed into a loudspeaker in the headphones.

Inverting the signal means the sound waves are opposite to the original background sounds. Adding opposite signals cancels out the noise, so the noise heard by anyone wearing noise-cancelling headphones is greatly reduced.

Music headphones and sensitive hearing aids are also available with this technology.

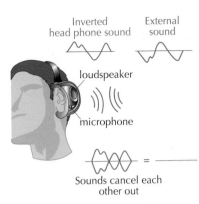

Figure 10.13 *Noise-cancelling headphones.*

GO! Activity 10.12

☻ **1.** Which part of the ear converts sound waves to electrical signals?

    A The cochlea   **B** The eardrum

    C The ear canal   **D** The auditory nerve

2. The loudness of sound can be measured using

    A a ruler   **B** a decibel meter

    C a voltmeter   **D** a stopwatch

3. Which of the following is not a way to protect your hearing?

    A Listening to quiet music

    B Standing closer to speakers

    C Reducing the time you listen to loud sounds

    D Wearing ear defenders

4. Which statement is not true?

    A Quiet sounds can harm human ears.

    B The sounds near to speakers can be louder than a road drill.

    C The eardrum can be damaged by playing music through your earphones.

    D A noisy factory can permanently damage the ears.

5. Which of the following jobs would not need hearing protection?

    A A librarian

    B A tree surgeon

    C An airport worker, working near aeroplane jet engines

    D A construction worker, using a pneumatic drill during road works

Uses of sound

Learning intentions

In this section you will:

- learn that there are different applications of sound waves
- investigate sound reproduction technologies.

Forces, electricity and waves: Vibrations and waves SCN 4-11a

Technology for recording, processing and reproducing sound is constantly being developed. Devices can generate and detect waves of different volumes and frequencies. Sound can be manipulated to change the quality and characteristics. There are many uses for sound waves, from medical and industrial applications, to the more obvious applications in entertainment and communications. Many applications use sounds that humans cannot hear. These are ultrasounds and infrasounds.

Ultrasound

Ultrasounds are high-frequency sounds above our hearing range (above about 20 000 Hz, or 20 kHz). Dog whistles work in the ultrasound region, producing sounds in the range 23–54 kHz, so they cannot be heard by humans, but can be heard by dogs and other animals.

Table 10.4 *Uses of ultrasound.*

Use of ultrasound	
Body scans	Ultrasound is used to scan body organs and unborn babies, allowing checks for anything unusual. A probe directs a high-frequency sound wave into the body. The sound wave reflects from different tissues in the body. The scanner detects the reflections and calculates distances knowing the speed of the wave and the time taken to receive the reflection. The information is displayed on a screen as an image.
Aircraft safety checks	Mechanics check for cracks in aircraft parts and metal pipes using ultrasound generators and detectors. When the ultrasound wave reaches a crack, part of the wave is reflected back. This can then be detected.
Fishing and ocean research	The distance to the sea floor or to shoals of fish can be measured using the echo from an ultrasound wave.
Breaking up kidney stones	High-powered ultrasound waves vibrate the kidney stone so it breaks apart. The pieces can then be passed out of the body.
Cleaning surgical equipment, electronic components, jewellery and teeth	Ultrasound waves produce bubbles in a liquid that clean the object.
Echolocation	Animals such as bats and dolphins use echolocation with ultrasound to detect prey and predators.

Infrasound

Infrasounds are low-frequency sounds below our hearing range (below about 20 Hz). These sounds transfer little energy, but can be detected using sensitive microphones.

Table 10.5 *Uses of infrasound.*

Use of infrasound	
Animal communication	Some large animals communicate using infrasound. Elephants and whales have large skulls which vibrate and amplify low-frequency sounds, directing them to the ear bones. This lets whales and elephants hear low-frequency sounds that we cannot hear.
Volcano detection	Scientists can detect infrasound from volcanoes that are about to erupt. Monitoring activity is vital to ensure early warning can be given to people living near active volcanoes.
Tracking meteors	As meteors enter the Earth's atmosphere they produce a shock wave that gives an infrasound wave. These waves can be detected from the ground.

Figure 10.14 *Elephants produce infrasound with their feet. These waves can carry several kilometres.*

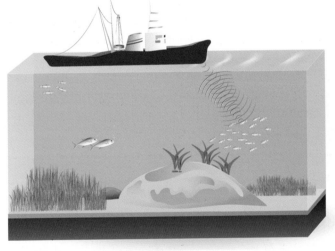

Figure 10.15 *Ultrasound is used to find shoals of fish. The distance is calculated knowing the speed of the wave and the time taken to receive the echo.*

> ### 🔍 Hint
>
> Remember that sound waves are not a form of radiation.

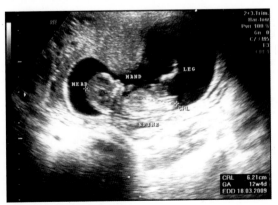

Figure 10.16 *An ultrasound scan of an unborn baby boy at 12 weeks.*

GO! Activity 10.13

☺ 1. Copy and complete the following passage using the words provided to describe how an ultrasound scan is used to detect an unborn baby.

frequency   reflects   calculates   ultrasound   probe   speed   time   screen

An _ _ _ _ _ _ _ _ _ _ machine transmits a high _ _ _ _ _ _ _ _ _ sound wave into the body using a _ _ _ _ _. The sound wave _ _ _ _ _ _ _ _ from the different tissues in the body. The machine detects these reflections and _ _ _ _ _ _ _ _ _ _ the distance knowing the _ _ _ _ _ of the wave and the _ _ _ _ taken to receive the reflection. This information is then displayed on a _ _ _ _ _ _ as an image.

2. Search online to research another use of ultrasound. This could be in industry or in medicine. How has this technology helped our lives? Write a short report on one way that sound technology helps us.

Sound reproduction

In the early days of sound recording and reproduction, sound was recorded using **analogue** techniques. The sound wave was directly recorded; the wave pattern was unaltered. This was the technology used for most of the 20th century, and was used for vinyl records and tape recordings.

In the 1980s, sound recording started using **digital** techniques. Sound waves are recorded and converted into digital electronic signals. This is the technology used for Compact Discs, mp3 files and most sound files you use on phones and computer equipment.

The use of digital processing electronics makes it much easier to change the recorded sounds and add effects. For example, Auto-Tune is a device which alters the pitch in musical recordings, so vocal tracks can be tuned to make the singer sound pitch-perfect. Digital processing also helps with applications such as voice recognition. You can get apps for your smartphone which will change the way your voice can sound when played back.

Digital sound technology also makes it easier to produce accurate copies of music.

Sound can also be processed and reproduced in hearing aids. Certain frequencies can be boosted to improve the quality of hearing for those who have hearing impairments.

📖 Word bank

- **Analogue signal**
Signals that have a continuous range of values are called analogue signals.

- **Digital signal**
Signals that have only two values are called digital signals.

Figure 10.17   *Different ways of listening to music.*

Recording and reproducing sound needs some essential equipment:

- **input**: a microphone (to collect sound waves and transfer them to electrical waves)

- **process**: sound processors and a storage device

- **output**: a loudspeaker.

Most microphones use an **electromagnet** that vibrates when the sound wave hits it. The vibrating electromagnet creates an electrical signal that represents the pattern of the sound wave.

Unidirectional microphones pick up sounds from one direction only. This is usually the type of microphone used by singers and performers onstage, because less background sound is detected. Omnidirectional microphones pick up sounds from all directions.

The electrical signal from the microphone is electronically processed and stored using a storage device such as a CD or a computer file. The signal can be manipulated before being sent to a loudspeaker.

The output of the storage device is sent to a loudspeaker. Loudspeakers use electromagnets to produce sound waves. Different types of loudspeaker are designed to reproduce different frequencies. For example, low-frequency sounds are produced using large 'woofers'. High-frequency sounds are produced using small 'tweeters'.

The electrical signal from the storage device is usually quite small, so loudspeakers usually need some kind of amplifier in order to generate sufficient volume. For small earphones such as those used with mp3 players and mobile phones, the electrical signal is big enough to power the earphones.

Make the link

Learn more about some of the equipment used in sound recordings in Chapter 7.

📖 Word bank

- **Electromagnet**

An electromagnet is a special type of magnet which has a metal core with a coil of wire around it.

🟢 Activity 10.14

☻ Audacity® is free, open source, cross-platform audio software for multi-track recording and editing. It can be used with the microphone built into your computer to analyse sound.

If possible, download this software or use another program to record and manipulate a sound. You might prefer to search online to find a free web-based audio editor.

Record your voice then apply an effect such as the echo. Listen to the manipulated sound.

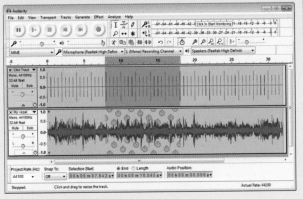

Audacity® software interface.

Learning checklist

After reading this chapter and completing the activities, I can:

N3 **L3** **N4** **L4**

- state that sound waves transfer energy. **Activity 10.1 Experiment 3** ○ ○ ○

- state that humans and animals can hear different frequency ranges. **Activity 10.2** ○ ○ ○

- explain what happens to the pitch and the waveform when the frequency of a sound wave is changed. **Activity 10.3, Activity 10.4** ○ ○ ○

- explain what happens to volume when the amplitude of a sound wave is changed. **Activity 10.3, Activity 10.4** ○ ○ ○

- describe what happens to a waveform if the frequency or amplitude are changed. **Activity 10.3, Activity 10.4, Activity 10.5** ○ ○ ○

- explain how to measure the speed of sound in air. **Activity 10.6, Activity 10.7** ○ ○ ○

- state that sound travels at different speeds in different materials. **Activity 10.7 Q1, Activity 10.8 Q1–Q4** ○ ○ ○

- use the relationship between speed, distance and time to calculate the speed of sound. **Activity 10.8 Q5–Q10** ○ ○ ○

- describe how to measure sound level using the decibel scale. **Activity 10.9** ○ ○ ○

- explain the term noise pollution. **Activity 10.11** ○ ○ ○

- describe the parts of the ear and what it can hear. **Activity 10.12 Q1** ○ ○ ○

- give examples of hearing protection. **Activity 10.12 Q3–Q5** ○ ○ ○

- describe some of the applications of sound waves. **Activity 10.13** ○ ○ ○

- describe different sound reproduction technologies. **Activity 10.14** ○ ○ ○

11 Light

You should already know:

- that light is reflected from shiny surfaces
- the order of the colours of a rainbow
- that we see objects because light rays enter our eyes from those objects
- mixing light of different colours produces other colours.

National 3

Light travels in straight lines

Learning intentions

In this section you will:

- learn that light travels in straight lines
- investigate applications based on the property that light travels in straight lines.

Light rays travel in straight lines. The light from the Sun travels in straight lines to the Earth as shown in the sunset in Figure 11.1(a).

When we look at a source of light, such as a candle or light bulb, we see the source because rays of light travel in straight lines directly into our eyes. **Ray diagrams** are used to represent light rays in drawings. Ray diagrams are drawn with a ruler and always have an **arrow** showing the direction of light, as shown in Figure 11.1(b).

(a)

(b)

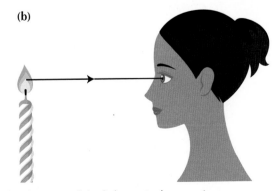

Figure 11.1 **(a)** *Light rays during a sunset.* **(b)** *The direction of the light ray is shown using an arrow.*

GO! Activity 11.1

😊😊 **1. Experiment** In this activity you will investigate the straight-line properties of light.

Read the plan carefully then follow the instructions, making sure to follow all of the safety rules as you would normally be expected to do when carrying out practical work.

You can work in small groups but you should record your own observations and complete the answers yourself.

You will need: three equally sized index cards, some sticky tack, a torch or ray box, a sharp pencil and a ruler.

Method

(a) On one card, use the ruler to draw diagonal lines connecting each opposite corner. Stack the three cards, with the diagonal lines on the drawn card facing up. Where the lines intersect, use the sharp pencil to create a hole in the centre of all three cards.

(b) Position each card vertically on the table at equal distances away, as shown in the figure, using sticky tack to hold them upright. Each card should be around 5–10 cm apart. The straight edge of the ruler can be used to ensure the three cards are aligned along one edge.

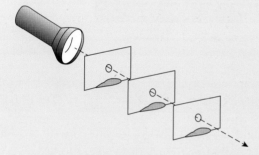

(c) Shine the torch through the hole in the first card. Direct it so that the light is pointing directly at the hole.

Can you see the light through the hole in the last card?

(d) Move the middle card slightly to the left or right. Is the light visible through the hole in the last card?

2. Explain how this experiment proves that light travels in straight lines.

🔍 Hint

You should be able to align the torch/ray box behind the first card so that the light passes through all three holes. Looking from in front of the third card, you can move the torch so that the light is directed toward the second hole and third hole.

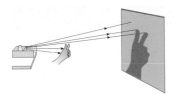

Figure 11.2 *Light rays are blocked, casting a shadow on a screen.*

Shadows

Shadows form when light rays are blocked by an **opaque** object. Light cannot travel around the object, so a dark area is created behind the object where there is no light. This is the shadow.

📖 Word bank

• **Opaque**

An opaque object is one that does not let light through it.

⁘ Make the link

Learn more about how shadows are used to explain eclipses in Chapter 17.

❓ Did you know ...?

Shadow puppetry originated over 2000 years ago in Southeast Asia, and is still popular today. A puppet is placed between a light and a screen. Moving the puppets can create the illusion that the figures on the screen are moving. Talented puppeteers can skilfully make the puppets walk, dance, fight or laugh.

Figure 11.3 *Shadow puppetry in Thailand.*

🔵 Activity 11.2

1. A ray diagram shows how light travels. A light ray must always be drawn with a ruler. State one other important feature of a light ray that is always drawn on the ray.

2. Draw a diagram to show how light travels from a television into our eyes.

3. Describe an experiment you could do to show that light travels in straight lines.

4. Draw a ray diagram to show how a shadow of a tree is formed by the light of the Sun.

5. In a darkened room, shine the light from a torch onto a wall, then put your hand in between the wall and the torch.

    (a) Explain why a shadow of your hand is created on the wall.

    (b) Move your hand closer to the torch. State and explain what happens to the size of the shadow on the wall.

Reflection of light

Learning intentions

In this section you will:

- investigate the law of reflection
- identify applications of reflection, including curved mirrors and optical fibres.

Most of the light that enters the eye does not come directly from a light source, like the Sun. We see most objects because light from the Sun or other light sources is reflected off the object into our eyes.

When light hits a rough object, the light reflects off in all directions. Most objects scatter light in this way. If the object is very smooth, such as a mirror, light is reflected more evenly.

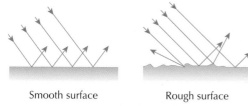

Smooth surface   Rough surface

Figure 11.5   *Light is reflected from a smooth surface evenly, and scattered unevenly from a rough surface.*

When light from an object is **incident** on a mirror, it is reflected evenly to create an **image**. The ray diagram in Figure 11.7 shows how an image is formed in a mirror. In this ray diagram, only two rays are drawn to keep it simple.

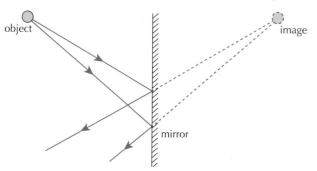

Figure 11.7   *The solid lines represent light rays. The dashed lines show where the light rays appear to be coming from – the image. This is the same distance away from the mirror as from the object to the mirror.*

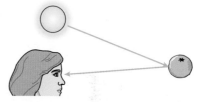

Figure 11.4   *Light from the Sun is reflected from the orange into the eye.*

Make the link

Compare the reflection of light off a smooth mirror with the reflection of sound from smooth and rough surfaces in Chapter 10.

Word bank

- **Incident**

When light hits an object, such as a mirror, the ray that travels toward the mirror is called the incident ray.

- **Image**

An image is the representation of an object when light is reflected in mirrors or when light passes through lenses.

Figure 11.6   *The reflection of light from a mirror allows you to see an image.*

Word bank

• **Plane mirror**

A plane mirror is a mirror with a flat, reflective surface. It is usually made of glass with a thin layer of highly polished aluminium or silver on the back.

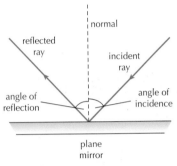

Figure 11.8 *The path of a ray of light reflected from a plane mirror.*

The law of reflection

When light hits a **plane mirror**, the light is reflected from the mirror in a predictable manner. The behaviour of light when reflecting off a plane mirror is known as the **law of reflection**.

For a plane mirror:
> **the angle of incidence = the angle of reflection**

Table 11.1 *A list of terms to describe reflection.*

Term	Description
Incident ray	The ray of light going to the mirror.
Reflected ray	The ray of light that comes from the mirror.
Normal line	The line drawn at right angles to the mirror. Light incident on this line would reflect back on the same path.
Angle of incidence	The angle measured between the incident ray and the normal line.
Angle of reflection	The angle measured between the reflected ray and the normal line.

Activity 11.3

1. Experiment In this activity you will test and verify the law of reflection.

Read the plan carefully then follow the instructions, making sure to follow all of the safety rules as you would normally be expected to do when carrying out practical work.

You can work in small groups but you should record your own observations and complete the answers yourself.

You will need: a ray box and single slit, a plane mirror and a protractor.

Method

(a) Set up the ray box to shine a single ray of light toward the plane mirror as shown.

(b) Set up the ray box to shine a single ray of light at an angle of incidence of 10° from the normal.

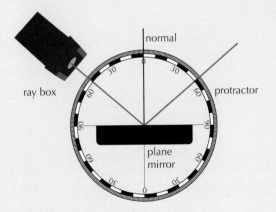

Ray box directed at plane mirror.

(c) Record the angle of reflection in a table.

Angle of incidence (°)	Angle of reflection (°)
10	
20	
30	
40	

> **Hint**
>
> Make sure the normal and the mirror are at 90° to each other, and that the ray hits the mirror where the normal line touches the mirror.

(d) Repeat with the other angles in the table and record each angle of reflection.

(e) How do the values for the angle of incidence and reflection compare? Can you use them to verify the law of reflection?

Applications of reflection

The image in a mirror is always upright, but it is also **laterally inverted** (swapped from one side to the other). When we look in a mirror directly, the light reflects directly off the mirror back along the normal line. It can be seen from Figure 11.9 that the left hand of the person is the right hand of the image. If you look at writing in a mirror, it will appear backwards, but the right way up.

Emergency services use the lateral inversion of images when designing signs on emergency vehicles such as ambulances and police cars.

> 📖 **Word bank**
>
> • **Lateral inversion**
>
> An image in a mirror is laterally inverted. This means the left- and right-hand sides of an image are reversed. Lateral means sideways.

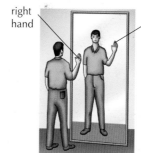

right hand · left hand of image

Figure 11.9 *Lateral inversion in a mirror.*

> **GO!** **Activity 11.4**
>
> 😊😊 **1. Experiment** In this activity you will attempt some laterally inverted writing.
>
> Read the plan carefully then follow the instructions, making sure to follow all of the safety rules as you would normally be expected to do when carrying out practical work.
>
> You can work in small groups but you should record your own observations and complete the answers yourself.
>
> *You will need: a mirror, some paper and a pencil.*
>
> *Method*
>
> **(a)** Look carefully at how the word 'Ambulance' is written in the figure below.
>
>

Figure 11.10 *The word 'Ambulance' is laterally inverted on the vehicle's bonnet so drivers can read it in their rear-view mirror.*

(continued)

(b) Write a word or message on a piece of paper that is laterally inverted.

(c) Give it to your partner to read with the mirror. Can they interpret the message correctly with the mirror?

(d) Repeat by allowing your partner to try the same with you.

2. Why are words laterally inverted on emergency service vehicles?

Word bank

- **Focus**

When light rays intersect they meet at a focus.

- **Converge**

When light rays are brought to a focus.

- **Diverge**

When light rays are spread out so that they move apart from one another.

Curved mirrors

Curved mirrors are used to bring rays of light to a sharp point (known as a **focus** or **focal point**) or to spread rays out.

Concave mirrors reflect light to a focus. The rays are said to **converge**.

Convex mirrors reflect light so that it spreads outward. The rays are said to **diverge**.

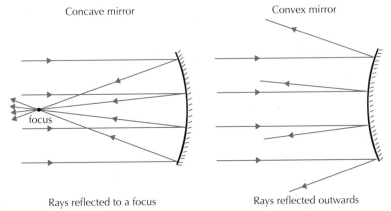

Concave mirror — focus — Rays reflected to a focus

Convex mirror — Rays reflected outwards

Figure 11.11 *Rays are reflected inward in concave mirrors and outward in convex mirrors.*

Figure 11.12 *A solar cooker used in the Himalaya mountains. Can you think of a reason why the pan is black?*

A solar furnace uses a concave mirror to reflect the Sun's rays to a focus. Concentrating solar energy with a mirror at one point can produce very high temperatures. In some parts of the world, small solar furnaces can be used as cookers. They are particularly suitable in remote regions in hot dry places where it is hard to find firewood. Large solar furnaces can generate electricity from the heat generated using large arrays of mirrors.

⟳ Keep up to date!

The world's largest solar furnace is in the Pyrenees mountains, on the French–Spanish border.

Figure 11.13 *The world's largest solar furnace in the Pyrenees–Orientales in France. Search online to find out how it works to harness the Sun's energy.*

Convex mirrors are often used for security and safety purposes. They help to see around blind corners or difficult road exits. They are commonly used in buses to increase security, allowing the driver to see into the top of a double-decker bus.

✴ Physics in action: Optical fibres

An optical fibre is a thin strand of glass, often with an outer plastic coating. Light is transmitted along fibres by the process of **total internal reflection** (TIR). Light rays travelling along the fibre are reflected off the inside surface

of the fibre, so even when the fibre is bent, the light still travels along the fibre.

Figure 11.16 *Light travelling down an optical fibre by total internal reflection.*

Optical fibres can be used to transmit light into places which cannot be lit using conventional lighting. Surgeons use optical fibres to see what they are operating on inside human bodies.

Optical fibres can be used to investigate the causes of blocked drains without having to dig up the whole drain.

Optical fibres can also be used to transmit information. Coded pulses of light are used to transmit huge quantities of information for internet traffic and for TV and telephone applications. Optical fibres can transmit much higher amounts of information down a single strand of glass compared with the older copper wires used in phone networks.

Figure 11.14 *Convex mirrors reflect light over a large angle, eliminating blind spots.*

Figure 11.15 *A convex mirror in a bus allows the whole deck to be viewed at once by the driver.*

(continued)

N3	L3	N4	L4

The improvement of broadband speeds in Britain has largely been the result of the increased use of fibre-optic cables for communication.

Figure 11.17 *A sign that fibre broadband is available in a rural area.*

GO! Activity 11.5

☺ 1. Copy and complete the following paragraph using the words provided:

reflection   **Sun**   **angle**   **equal**   **eye**   **light**

When we look at an apple on a table, the _ _ _ _ _ from the _ _ _ is reflected into my _ _ _ . The law of reflection states that the _ _ _ _ _ of incidence is _ _ _ _ _ to the angle of _ _ _ _ _ _ _ _ _ _ .

2. The diagram below shows a ray of light incident on a mirror.

   (a) Complete the diagram, adding the following labels: angle of incidence, angle of reflection, normal line, incident ray, reflected ray.

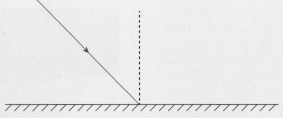

   (b) The angle of incidence is 45°. What is the angle of reflection?

3. Copy and complete the following diagrams to show how light is reflected off a concave and convex mirror.

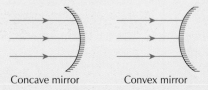

Concave mirror   Convex mirror

4. Describe an application of a concave mirror.

5. Describe an application of a convex mirror.

6. Copy and complete the diagram below to show how light reflects along an optical fibre.

light beam

7. Search online to find out some advantages of using optical fibre instead of copper wires for the transmission of information.

Lenses and refraction of light

Learning intentions

In this section you will:

- learn that refraction is the change of direction of light as it travels from one material to another
- investigate the refraction of light when it passes through different materials, for example lenses and prisms
- learn how light changes direction in convex and concave lenses
- identify applications of refraction, such as lenses to correct long and short sight.

National 3

Curriculum level 3

Forces, electricity and waves: Vibrations and waves SCN 3-11a

National 4

(a)

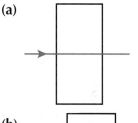

(b)

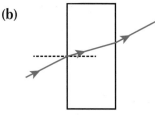

Figure 11.18 **(a)** *Ray of light incident on glass along normal line.* **(b)** *Ray of light incident on glass at an angle to the normal line.*

The diagrams in Figure 11.18 show what happens when light passes through a block of glass. The ray in Figure 11.18 (a) is incident along the normal line, with an angle of incidence of 0°. The ray in Figure 11.18 (b) is incident at an angle greater than 0° to the normal, and **refraction** happens.

When light travels from one medium to another, it can change direction. This is due to a change in speed. This change of direction only happens when the angle of incidence is greater than 0°.

❓ Did you know ...?

When a pencil is placed in a glass of water, it appears to bend. This is due to the light **refracting** as it leaves the water.

Figure 11.19 *The light bends as it leaves the water, causing the pencil to appear bent.*

📖 Word bank

- **Refraction**

When light enters a medium at an angle greater than 0°, it changes direction. This is known as refraction.

GO! Activity 11.6

😊😊 **1. Experiment** In this activity you will investigate how light changes direction in two differently shaped pieces of glass: a glass block and a triangular prism.

Read the plan carefully then follow the instructions, making sure to follow all of the safety rules as you would normally be expected to do when carrying out practical work.

You can work in small groups but you should record your own observations and complete the answers yourself.

You will need: a ray box, a rectangular glass block, a triangular glass prism, white A4 paper, a ruler, a protractor.

Method

(a) Place a glass block in the centre of a piece of white paper. Draw around it carefully to show the outline, then lift the block and draw a normal line at the mid-point of the long edge of the glass block. Use a protractor and ruler.

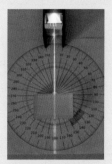

Figure 11.20

(b) Put the glass block back on the outline on the white paper. Set up the ray box to shine a single ray of light into the glass block, along the normal line, as shown in Figure 11.20.

(c) Increase the angle of incidence with the glass by moving the ray box clockwise around the block, as shown in Figure 11.21. What happens to the ray of light in the block?

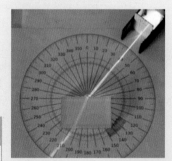

(d) How does the angle of incidence of the light (between the ray and the normal line) compare with the angle of the refracted light as it leaves the glass block?

> 🔍 **Hint**
>
> Use a protractor to measure the angle between the normal and the incident ray, and the normal and the refracted ray that leaves the block of glass.

Figure 11.21

(e) Remove the rectangular block and turn the page over. Place the triangular prism on the page, triangle side down, as in Figure 11.22. Shine the light from the ray box toward one side of the prism, at 90° to the surface of the prism (along the normal line). Turn the prism slowly clockwise so that the incident ray is not along the normal line. Draw a diagram to show what happens to the ray of light as it passes through the prism.

Figure 11.22

N3 L3 N4 L4

Convex and concave lenses

Lenses use refraction to bring light to a focal point, or spread it out.

When parallel rays of light pass through a **convex** lens, the rays are brought into focus at the **focal point**, as shown in Figure 11.23.

A **concave** lens takes parallel rays of light and spreads them out, as shown in Figure 11.24.

📖 Word bank

• **Focal point**

The focal point of a lens is where the refracted rays meet at a point.

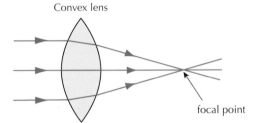

Convex lens

focal point

Figure 11.23 *Light rays are brought to a focus with a convex lens.*

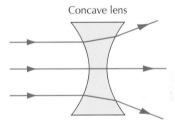

Concave lens

Figure 11.24 *Light rays are spread out with a concave lens.*

GO! Activity 11.7

 1. Experiment In this activity you will investigate how light refracts as it passes through different convex and concave lenses.

Read the plan carefully then follow the instructions, making sure to follow all of the safety rules as you would normally be expected to do when carrying out practical work.

You can work in small groups but you should record your own observations and complete the answers yourself.

You will need: a ray box with three slits, one thick convex lens, one thin convex lens, one concave lens and some blank paper.

Method

(a) Place a piece of blank paper on the desk. Set the ray box and the thin convex lens on the paper, so that the three rays of parallel light are directed into the thin convex lens. Carefully draw around the lens and carefully mark the path of the rays of light on both sides of the lens.

Rays of light shining on a thin convex lens.

(b) Without moving the ray box, replace the lens with the thicker convex lens. What do you notice has happened to the refracted rays due to this lens? Has the focal point moved?

(continued)

(c) Summarise how the thickness of the lenses affects the paths of the rays of light and the location of the focal point.

(d) Now repeat step (b) on a new sheet of paper using the concave lenses. How does the light behave when it passes through a concave lens?

Rays of light shining on a concave lens.

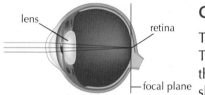

Figure 11.25 *Normal vision.*

? Did you know ...?

The image formed on the retina in our eye is upside down. Our brain has to decode the signal sent by the optic nerve, flipping it upright.

Correcting sight defects

The eye works by bringing rays of light to a focus using a lens. The lens at the front of the eye focuses light on to the retina at the back of the eye. If an object is far away from the eye, the shape of the lens is thin, allowing the light to be focused on the retina. When an observer then looks at a close-up object, the lens changes shape. It becomes thicker, to allow the light to be focused correctly on the retina.

If the lens does not work correctly, people may suffer from short- or long-sightedness defects.

Short-sightedness is very common. A person with short sight cannot see objects that are far away from them very clearly. This is because the light travelling through the lens in the eye focuses short of the retina at the back of the eye.

Short sight can be corrected by spreading the rays out before they reach the lens in the eye by adding a concave lens.

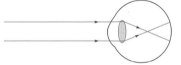

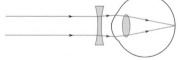

Figure 11.26 *The light lands short of the retina for a short-sighted person.*

Figure 11.27 *The concave lens corrects the vision by moving the focal point to the retina.*

Long-sightedness is less common than short sight. A person with long sight cannot focus on objects that are close to them. The light travelling through the lens in the eye focuses past the retina. The most common sign of long-sightedness is difficulty in reading books and magazines.

The likelihood of becoming long-sighted increases with age. Many people with perfect vision when they are young find they need reading glasses as they move into their 50s.

Figure 11.28   *The eye lens becomes stiffer with age, causing long sight for many as they get older.*

Long sight can be corrected by bringing the rays inward with a convex lens placed in front of the eye.

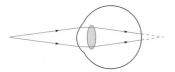

Figure 11.29   *The light lands long of the retina for a long-sighted person.*

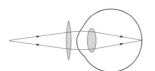

Figure 11.30   *A convex lens corrects the vision by bringing the focal point back to the retina.*

★ Memory Aid

Remember the difference between short- and long-sightedness:

- Light focuses **short** of the retina for a **short-sighted** person.
- Light focuses **long** of the retina for a **long-sighted** person.

GO! Activity 11.8

☺ 1. State what is meant by 'refraction'.

2. Consider the diagram of a lens on the right.
   (a) What is the name given to this type of lens?
   (b) Copy and complete the diagram showing what happens to the rays of light when they leave the lens.
   (c) What common eye defect can be corrected by this type of lens?

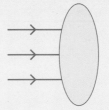

3. Consider the diagram of a lens on the right.
   (a) What is the name given to this type of lens?
   (b) Copy and complete the diagram showing what happens to the rays of light when they leave the lens.
   (c) What common eye defect can be corrected by this type of lens?

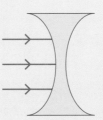

(continued)

N3   L3   N4   L4

4. Consider the three convex lenses shown in the diagram below.

A   B   C

When parallel rays of light are incident on each lens in turn,

(a) which of the above lenses focuses the light closest to the lens?

(b) which of the above lenses focuses the light farthest from the lens?

(c) Describe an experiment that you could use to confirm your answers to (a) and (b).

5. Look at the diagram of an eye:

(a) What is the name given to this common eye defect?

(b) What type of lens can be used to correct this problem?

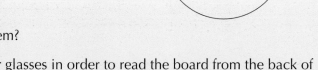

6. A pupil needs to wear glasses in order to read the board from the back of the classroom. However, she can read the textbook without glasses.

(a) What type of sight defect does this student have?

(b) What type of lens would you use to correct this sight defect?

Colour

Learning intentions

In this section you will:

- learn that a visible spectrum is formed when light disperses in a prism
- investigate how mixing colours can produce white light and other colours
- learn how coloured light affects the appearance of different coloured objects
- investigate different sources of light with a spectrometer
- describe colour blindness.

When white light passes through a triangular glass prism, a **spectrum** of colours is formed.

White light is made up of all possible colours of light. When white light passes through a triangular prism, each colour is refracted by a slightly different amount. This is because the colours of light have different wavelengths. When the light leaves the prism each colour is **dispersed**.

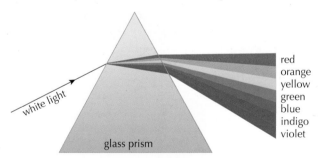

Figure 11.32 *The dispersion of white light into a coloured spectrum by a prism. Red light is refracted least and violet light is refracted most.*

The spectrum is commonly described as having seven colours:

red, orange, yellow, green, blue, indigo and violet.

Sir Isaac Newton carried out experiments in the seventeenth century to show that white light is made up of the colours of the spectrum. Although many other colours can be distinguished, reducing the number to seven makes it easier to remember the order.

📖 Word bank

- **Spectrum**

The visible spectrum is the range of coloured light our eyes can detect.

Figure 11.31 *A spectrum of colours produced by a prism.*

📖 Word bank

- **Dispersion**

Dispersion is the separation of light into colours by refraction, resulting in the formation of a spectrum.

⚗️ Make the link

The wavelengths of visible light are explored in Chapter 12.

★ Memory Aid

Remember the colours of the spectrum using this phrase:

Richard Of York Gave Battle In Vain

where the first letter of each word in the phrase represents a colour.

Another helpful method of remembering the colours is with the abbreviation ROY G BIV.

? Did you know ...?

Rainbows are formed when the white light from the Sun is refracted in water droplets in the air. The light is reflected inside the drop, refracted at the edge of the drop and dispersed when it leaves the drop.

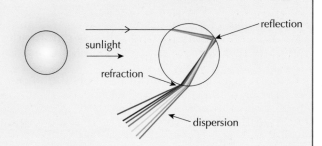

Figure 11.33 *A rainbow formed over the Forth Rail Bridge.*

Figure 11.34 *Reflection, refraction and dispersion all take place in a drop of water to form a rainbow.*

📖 Word bank

- **Primary colours**

The primary colours of light are red, blue and green.

Mixing coloured light

The three **primary colours** of light are red, blue and green. Mixing these colours of light together produces white light.

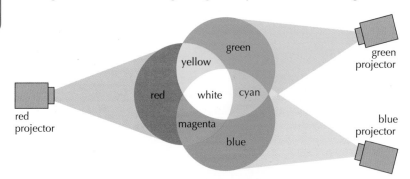

Figure 11.35 *Mixing the primary colours of light.*

📖 Word bank

- **Secondary colours**

Secondary colours are the colours produced when primary colours are mixed.

Figure 11.35 shows how the three primary colours mix together to form white light. Mixing two primary colours together produces **secondary colours**. Table 11.2 shows the secondary colours produced when two primary colours are mixed.

Table 11.2 *Primary and secondary colours*

Primary colours mixed	Secondary colour produced
Red and blue	Magenta
Red and green	Yellow
Blue and green	Cyan

🔵GO! Activity 11.9

😊
😊😊 **1. Experiment** In this activity, you will investigate how the primary colours can be mixed together to produce white light and the secondary colours.

Read the plan carefully then follow the instructions, making sure to follow all of the safety rules as you would normally be expected to do when carrying out practical work.

This practical would normally be demonstrated by your teacher.

You will need: colour mixing apparatus similar to that shown in the diagram. Separate red, green and blue LEDs connected to a battery can also be used.

Method

(a) Set up the three LED lights so that each light mixes on the screen to form a secondary colour, and all three also overlap to produce white light. Then turn off the LEDs.

(b) Turn on the following pairs of LEDs and record the colour formed when they mix.

   (i) Red and green

   (ii) Red and blue

   (iii) Blue and green

The colour mixing kit.

(c) Turn on all three LEDs. What colour of light is formed from mixing all three lights?

(d) Adjust the brightness of each coloured LED in turn. What change can be seen to each secondary colour?

🔍 Summary

The three primary colours mix together to form white light. Adjusting the brightness of each primary colour alters the secondary colour.

☀️ Physics in action: Television displays

Television sets have changed considerably over the past two decades. Before the turn of the century, most TV screens were large and very heavy.

By 2007 however, more flat-screen TVs with LCD (Liquid Crystal Display) screens were sold. Today, most televisions sold are now very thin and flat, and many use LEDs (Light Emitting Diodes). The displays on smartphones and tablets also use LEDs or LCD screens.

What has not changed is how the colours are produced on the displays. All displays, ranging from the first colour televisions of the twentieth century to the very latest smartphone, produce colour by mixing the three primary colours: red, blue and green.

Figure 11.36 *Displays have become thinner and lighter.*

Coloured materials

When white light from the Sun or a lamp reflects off different coloured objects, why do they all not look white? Each object appears a different colour, even though they are all illuminated with white light. This is because certain colours in the white light are absorbed by the object, while other colours are reflected. For example, green leaves are green because the surface of the leaves reflects the green light, and absorbs all the other light.

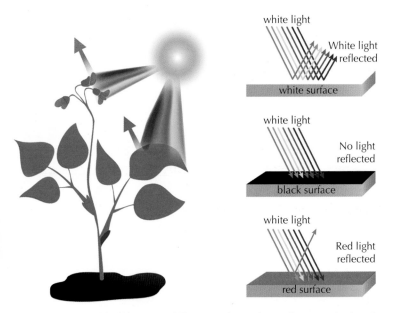

Figure 11.38 *Objects appear different colours depending on which colours of light are absorbed or reflected.*

Many common sources of light (such as the Sun and filament lamps) produce white light. Some light sources produce light of one colour, such as lasers and LEDs, which produce single colours. Coloured filters can also be used with a white light source to produce coloured light.

When light from a blue light source is directed toward a red object, the object will appear black. All the red light is absorbed, and no light is reflected, making it appear black.

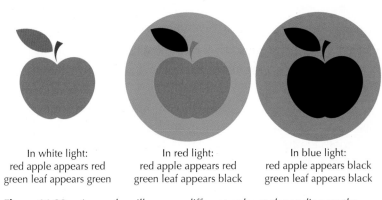

In white light:
red apple appears red
green leaf appears green

In red light:
red apple appears red
green leaf appears black

In blue light:
red apple appears black
green leaf appears black

Figure 11.39 *An apple will appear different colours depending on the colour of the light illuminating it.*

N3 L3 N4 L4

GO! Activity 11.10

😊
😊😊
1. Experiment In this activity you will investigate how coloured clothing appears in different coloured light.

Read the plan carefully then follow the instructions, making sure to follow all of the safety rules as you would normally be expected to do when carrying out practical work.

You can work in small groups but you should record your own observations and complete the answers yourself.

You will need: various items of clothing of different colours, different white light sources (e.g. a filament lamp, a candle, a 'daylight coloured' LED lamp), different coloured lamps (a white light with coloured filters can be used).

Method

(a) Arrange the clothing on your desk in a darkened room. Look at the clothing with each white light source in turn. Describe the effect of the different types of white light on the appearance of the clothing.

(b) Now shine different coloured lights onto the clothing. Describe the effect of the different types of coloured light on the appearance of the clothing.

2. Why do you think it is important to consider the colour of the light when trying on new clothes?

White light sources

A white light source is simply something that produces white light. The Sun, incandescent filament lamps, energy-saving fluorescent lights and LED lamps are all examples of white light sources.

Figure 11.40 *Four white light sources: a candle, an incandescent lamp, a fluorescent lamp and an LED lamp.*

We usually say that white light is made up of all the colours of the visible spectrum. However, not all white light sources look the same. This is because a white light source may not include all the colours of the visible spectrum in equal amounts. Some white light sources may include more of one colour than another.

A device called a **spectrometer** can be used to investigate the colours of light coming from different white light sources. A spectrometer allows the light to be split up to examine how much of each colour is present in the light.

Figure 11.41 shows how the spectrum of colours produced by different white light sources can vary.

  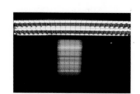

Incandescent lamp Energy-saving lamp Fluorescent tube lighting

Figure 11.41 *Light spectra from white light sources*

| N3 | L3 | N4 | L4 |

Fluorescent energy-saving lamps are part of a family of lamps called **gas discharge lamps**. They work by sending an electrical charge through a highly energised gas. Gas discharge lamps are used for neon lighting and other specialist lighting purposes.

Figure 11.42 *A gas discharge lamp in a school lab. The lamp uses krypton gas, which produces a 'bluish' white coloured light.*

White LED lighting for the home comes in two main types, warm or cool. Warm lighting is closer to traditional filament lighting. Cool LED lights produce a colour of white light that resembles daylight.

Figure 11.43 *Cool and warm LED lighting.*

Activity 11.11

 1. **Experiment** In this activity, you will make a spectrometer using an old CD to investigate the spectrums of colour produced by various white light sources.

Read the plan carefully then follow the instructions, making sure to follow all of the safety rules as you would normally be expected to do when carrying out practical work.

You can work in small groups but you should record your own observations and complete the answers yourself.

You will need: an old CD, a small cardboard box (a small shoebox would do), scissors or a craft knife, white light sources (incandescent lamp, fluorescent lamp tube, gas discharge lamp).

Method

(a) Create an eyehole by cutting a small hole in the top of the cardboard box. Create a slit under this eyehole for the CD to slot into at an angle of approximately 60°. On the other side of the box, opposite the CD, cut another thin slit to allow the light from the light source to enter.

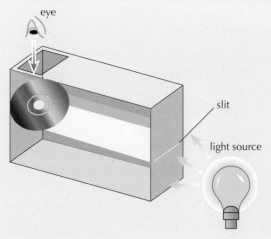

A CD spectrometer.

(b) Set up the lamp sources as shown above.

 Lamps can become hot. Be careful not to touch them.

Hint

If you cannot source a suitable cardboard box, search online to find a CD spectrometer template.

(c) Hold the spectrometer so that light can illuminate the CD through the slit. Look down at the spectrum formed from each light source.

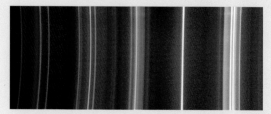

Spectrum seen through a CD spectrometer of a krypton gas discharge lamp.

(d) Draw the spectrum that you see for each white light source. Describe how they differ from each other.

GO! Activity 11.12

☺ **1.** Copy and complete the diagram below to show the colours of the spectrum produced when white light passes through a prism.

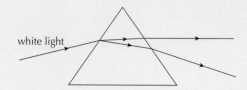

white light

2. Copy and complete the following passage using the words provided.

reflected   spectrum   absorbs   white light   colours

A blue jumper appears blue under _ _ _ _ _ _ _ _ _ _ because it _ _ _ _ _ _ _ all the _ _ _ _ _ _ _ of the _ _ _ _ _ _ _ _ except blue. The blue light is _ _ _ _ _ _ _ _ _.

3. Red light is shone on a blue jumper. What colour will it appear?

4. White light shines on a yellow daffodil. Explain why it appears yellow.

5. Red, green and blue light is mixed together. What colour of light is produced?

6. Explain what is meant by colour blindness.

7. Search online to try an online colour blindness test to determine how well you can see colours.

National 3

Curriculum level 3

Forces, electricity and waves: Vibrations and waves SCN 3-11a

Optical instruments

Learning intentions

In this section you will:

- learn how a periscope works
- learn how a pinhole camera works
- describe how the operation of the eye compares to a simple camera
- learn how telescopes and microscopes use lenses to magnify objects
- identify and describe commercial, medical and industrial uses of optical instruments.

There are many optical instruments that make use of reflection and refraction, such as:

- periscopes
- cameras
- telescopes
- microscopes.

The periscope

A periscope uses the **law of reflection** to allow observation over or around an obstacle.

Two mirrors are placed at 45° angles at the top and bottom of a rectangular column. Light entering at the top is reflected off each mirror to the observer at the bottom.

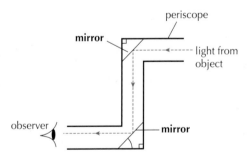

Figure 11.45 *Light reflects off both mirrors to see over an obstacle.*

? Did you know ...?

Periscopes were used extensively during both the First and Second World Wars. They were used in the trenches and in submarines. Today optical periscopes in submarines have been replaced with electronic imaging sensors.

Figure 11.44 *A British soldier looks towards the enemy lines using a periscope. Belgium, 1914.*

Cameras

A pinhole camera is a simple camera without a lens. Light passes through a small hole at the front of the camera, forming an image at the back of the camera.

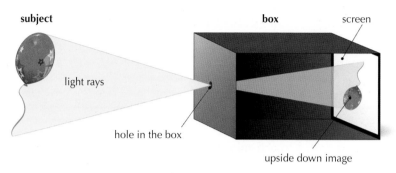

Figure 11.46 *How an image is formed with a pinhole camera.*

(GO!) Activity 11.13

😊😊😊 **1. Experiment** In this activity you can investigate how an image is formed using a pinhole camera.

Read the plan carefully then follow the instructions, making sure to follow all of the safety rules as you would normally be expected to do when carrying out practical work.

You can work in small groups but you should record your own observations and complete the answers yourself.

You will need: a cardboard box with a screen at the back made of greaseproof paper and a large hole at the front, black paper, a pin, a wide-coiled filament lamp.

Method

(a) Place a piece of black paper at the front of the camera and put a small pinhole in it. Point it at the filament lamp and observe the image on the screen. Draw the image.

Pinhole camera set-up.

(continued)

(b) How does the image appear compared to the lamp itself?

(c) Move the pinhole camera closer to the lamp. What happens to the image on the screen?

(d) Now add a few more holes to the black paper at the front of the camera. What happens to the image on the screen?

(e) Make one of the holes bigger (pencil sized). What effect does this have on the image on the screen?

The filament lamp shown produces an image like this with the pinhole camera.

A camera works in a similar way to the eye. Light enters through a hole at the front, passes through a lens and forms an image. The image is upside down on both the retina in the eye, and the sensor (or film) in a camera.

One key difference in the operation of a camera and the eye is the process of focusing on near and distant objects.

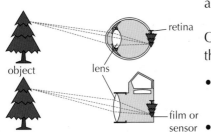

Figure 11.47 *Images are formed in the same way in a camera and the eye.*

- In the eye, the lens stays in the same place, but the shape of the lens is changed by muscles around the lens.

- In a camera, the lens stays the same shape but it is moved closer to or further away from the film or sensor to focus on objects at different distances.

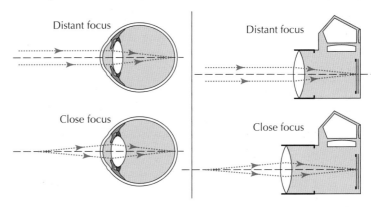

Figure 11.48 *How the eye and camera differ when focusing on objects at different distances away.*

Telescopes and microscopes

Telescopes and microscopes use lenses to magnify objects. Telescopes are used to magnify objects that are far away. Microscopes allow very small objects to appear larger.

A refracting telescope has two lenses. The large one at the front, called the **objective lens**, collects the light from distant objects. The smaller one, called the **eyepiece lens**, magnifies the collected light.

Figure 11.50 *Telescopes are used to study the universe.*

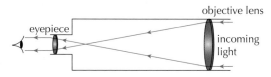

Figure 11.49 *The refracting telescope.*

⏵ Activity 11.14

1. Experiment In this activity you will make a simple refracting telescope.

⚠ Read the plan carefully then follow the instructions, making sure to follow all of the safety rules as you would normally be expected to do when carrying out practical work. Do not look directly at the Sun through any of the lenses you use, or through the telescope you make.

You can work in small groups but you should record your own observations and complete the answers yourself.

You will need: a metre stick, an objective and eyepiece lens, lens holders and a white card.

Method

(a) For each lens, hold it up between a window and the piece of card. By adjusting the distance between the lens and the card, observe when a sharp image of the window is formed on the card. Measure and record the distance between the card and the lens in each case. This is the focal length of the lens.

Once a clear image is formed on the card, measure the distance between the image and the lens with a ruler.

> 🔍 **Hint**
>
> The card will need to be a few centimetres from the lens to form an image with the smaller eyepiece lens. The larger objective lens will need to be much further from the card (perhaps 30–40 cm) before a sharp image is formed.

(continued)

| N3 | L3 | N4 | L4 |

(b) Place each lens in the lens holder and attach to the metre rule. The distance between the lenses must be equal to the sum of the two distances measured in (a) This forms the basis of the telescope.

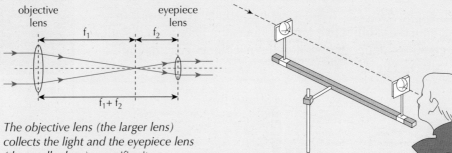

objective lens   eyepiece lens

f_1 f_2

$f_1 + f_2$

The objective lens (the larger lens) collects the light and the eyepiece lens (the smaller lens) magnifies it.

Lenses attached to a metre stick for a simple optical telescope.

(c) Direct the telescope at a distant object and look through the eyepiece. Some guidance may be required to see the magnified image.

2. If your school has access to a commercial refracting telescope, use it to view the same object from a similar distance. How does the image seen with your telescope compare with the image produced by the commercial telescope?

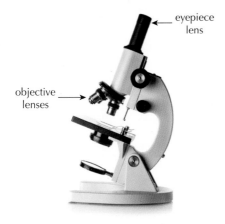

eyepiece lens

objective lenses

Figure 11.51 *An optical microscope.*

The microscope

A microscope works in much the same way as a telescope. An objective lens is used for collecting the light from the object, and bringing it to a focus. The objective lens on a microscope is much smaller than a telescope's lens, and is much more spherical, bringing the rays to a focus near the lens. The eyepiece lens magnifies the light from the objective lens.

Microscopes commonly have more than one objective lens. The thicker the objective lens is, the higher the **magnification**. Optical microscopes can be used to look at objects as small as one-thousandth of a millimetre.

📖 Word bank

• **Magnification**

The magnification of a lens refers to the number of times bigger the image is compared to the actual size of the object.

GO! Activity 11.15

☻ 1. The periscope in this figure is incomplete.

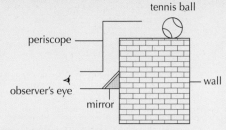

(a) What is missing from the periscope?

(b) Copy and complete the diagram to show how the periscope allows the ball to be seen.

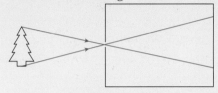

(c) What property of light allows the image of the object to be visible through the periscope?

2. Describe two similarities between the eye and an optical camera.

3. Describe one difference between the eye and an optical camera.

4. A pinhole camera is pointed at a Christmas tree, as shown. Two rays of light show how an image is formed of the Christmas tree by the pinhole camera.

(a) Copy the diagram and complete it by showing how the image would appear at the back of the pinhole camera.

(b) The hole is made bigger. What effect does this have on the image?

5. Telescopes and microscopes have dramatically improved our understanding of the universe. Search online to find out two discoveries made by:

(a) the refracting telescope

(b) the microscope.

Learning checklist

After reading this chapter and completing the activities, I can:

N3 L3 N4 L4

- state that light travels in straight lines. **Activity 11.1, Activity 11.2 Q1–Q3**   ○ ○ ○

- explain that shadows are formed because an object blocks the light reaching a surface because the light cannot travel around it. **Activity 11.2 Q4, Q5**   ○ ○ ○

- state that when light reflects off a plane mirror, the angle of incidence is equal to the angle of reflection. **Activity 11.3, Activity 11.5 Q2**   ○ ○ ○

- state that we see objects when light reflects off them into our eyes. **Activity 11.5 Q1**   ○ ○ ○

- state that light comes to a focus when it is reflected off a concave reflector. **Activity 11.5 Q3**   ○ ○ ○

- state that light spreads out when it is reflected off a convex reflector. **Activity 11.5 Q3**   ○ ○ ○

- give examples of applications of reflection of light from a curved mirror. **Activity 11.5 Q4, Q5**   ○ ○ ○

- state that light travels through an optical fibre by total internal reflection. **Activity 11.5 Q6**   ○ ○ ○

- describe how light is refracted in a rectangular glass block and a glass prism. **Activity 11.6**   ○ ○ ○

- state that refraction is the changing direction of light when it passes from one material to another. **Activity 11.8 Q1**   ○ ○ ○

N3 L3 N4 L4

- state that light is refracted to a focus with a convex lens. **Activity 11.7, Activity 11.8 Q2** ○ ○ ○

- state that light is refracted outward with a concave lens. **Activity 11.7, Activity 11.8 Q3** ○ ○ ○

- state that thicker convex lenses bring light to a focus closer to the lens. **Activity 11.8 Q4** ○ ○ ○

- explain how convex and concave lenses are used to correct sight defects. **Activity 11.8 Q5, Q6** ○ ○ ○

- state that white light can be formed by mixing red, green and blue coloured lights. **Activity 11.9, Activity 11.12 Q5** ○ ○ ○

- state that a visible spectrum is formed when white light is dispersed by a prism. **Activity 11.12 Q1** ○ ○ ○

- explain why objects appear different colours when illuminated by different coloured lighting. **Activity 11.10, Activity 11.12 Q2–Q4** ○ ○ ○

- identify white light sources such as sunlight and incandescent, energy-saving and discharge lamps using a spectrometer. **Activity 11.11** ○ ○ ○

- explain what is meant by `colour blindness'. **Activity 11.12 Q6, Q7** ○ ○ ○

- name and identify various optical instruments, including periscopes, pinhole cameras, lens cameras, telescopes and microscopes. **Activity 11.13, Activity 11.14, Activity 11.15** ○ ○ ○

- describe how the operation of the eye compares to that of a simple camera. **Activity 11.15 Q2, Q3** ○ ○ ○

- name and describe commercial, medical and industrial uses of optical instruments. **Activity 11.15 Q5** ○ ○ ○

12 EM spectrum

The electromagnetic spectrum

National 3

Curriculum level 3

Forces, electricity and waves: Vibrations and waves SCN 3-11b

National 4

Curriculum level 4

Forces, electricity and waves: Vibrations and waves SCN 4-11b

Learning intentions

In this section you will:

- identify parts of the electromagnetic spectrum
- give examples of applications associated with each type of electromagnetic radiation
- describe limitations associated with electromagnetic radiation beyond the visible spectrum
- state hazards associated with each type of electromagnetic radiation
- state approaches to minimising risk for hazardous regions of the electromagnetic spectrum
- identify how the invisible parts of the electromagnetic spectrum can be detected
- describe how electromagnetic radiation has impacted upon society and our quality of life.

 **Make the link**

You can find out more about sound waves in Chapter 10.

Just as sound travels through the air as a wave, visible light travels through space as a wave. It forms part of a range of radiations known as the **electromagnetic (EM) spectrum**. The table shows some differences between sound waves and EM waves.

Table 12.1 *Sound waves and EM waves*

	Sound waves	Electromagnetic waves
	Cannot travel through a **vacuum**	Can travel through a **vacuum**
Frequency range	20 Hz to 20 kHz (human hearing range)	100 to 10^{22} Hz
Wavelength range	1.7 cm to 17 m	10^{-14} to 10^6 m
Speed	340 m/s in air	300 000 000 m/s in air (880 thousand times faster than sound)

📖 **Word bank**

- **Vacuum**

A vacuum is an area of space that contains no matter particles at all.

Like sound waves, the frequency and wavelengths of electromagnetic waves vary. Different frequency ranges have different properties and different names. The EM spectrum is shown in Figure 12.1.

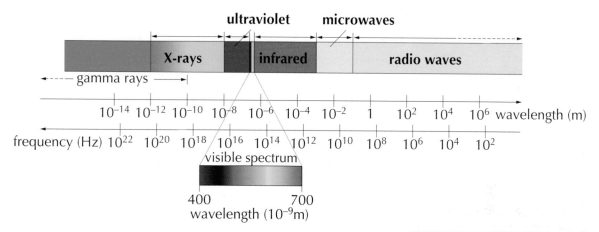

Figure 12.1 *The different parts of the electromagnetic spectrum.*

Most of the EM spectrum is invisible to human eyes. The small part of the EM spectrum that can be seen is the **visible spectrum**.

Different parts of the EM spectrum can be used over a wide range of applications. However, they can also present a range of risks to human health, and there are safety measures which need to be taken when using EM radiation. Other types of detection equipment are needed for the different parts of the EM spectrum.

The different properties of different parts of the EM spectrum are all determined by the wavelength and frequency of the EM waves.

Gamma rays

Gamma rays have the highest frequency of all EM waves and have the greatest energy of all types of EM radiation. Gamma rays are emitted from the nuclei of unstable atoms.

Table 12.2 *Frequency and wavelength range of gamma rays*

Gamma rays	
Frequency range	Greater than 10^{20} Hz
Wavelength range	Less than 10^{-12} m

Applications include the use of gamma rays as radioactive tracers to diagnose illness. Gamma rays also cause damage to human body cells. This property can be used in a range of

Make the link

The speed of any part of the electromagnetic spectrum can be determined using the equation:

$v = f\lambda$

where

v is the speed
f is the frequency
λ is the wavelength. λ is the Greek letter lambda.

This is the same equation described in Chapter 9 to determine the speed of water waves and sound waves. Since all types of electromagnetic radiation are waves, the same wave equation can be used.

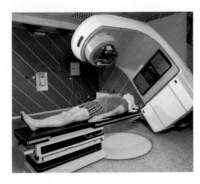

Figure 12.2 *Gamma rays emitted from this machine target cancer cells. Exposure to healthy cells is reduced by rotating the machine around the body.*

medical applications, but it also requires precautions to be taken when using gamma rays. Gamma rays can be used for:

* killing cells inside the human body

* sterilising medical instruments

* killing harmful bacteria on food.

Gamma rays are detected with a Geiger counter or a gamma camera.

X-rays

X-rays have a lower frequency and lower energy than gamma rays.

Table 12.3 *Frequency and wavelength range of X-rays*

X-rays	
Frequency range	Between 3×10^{16} and 10^{20} Hz
Wavelength range	Between 10^{-8} and 10^{-12} m

The most common application of X-rays is for viewing inside the body without the need for surgery. X-rays pass through the soft tissue in the body, but are absorbed by the bone. X-rays are detected by photographic film so photographic images of fractures and other injuries can be produced.

Other applications of X-rays include:

* airport security scanners

* inspection of welded joints in metal structures

* analysis of what lies under the surface of old paintings.

Airport security scanners use X-rays to examine the contents of luggage for dangerous objects, such as weapons or bombs.

> ### Make the link
>
> The properties of gamma radiation are explained in Chapter 13.

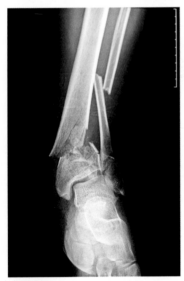

Figure 12.3 *An X-ray of a leg fracture.*

Figure 12.4 *An X-ray scan of a bag containing weapons.*

X-rays have lower energy than gamma rays, but can still cause damage to healthy tissue and cells. For this reason, **radiographers** who administer X-rays to patients take safety precautions:

- leaving the room or standing behind a lead-lined screen

- wearing lead aprons which block X-rays from passing through to the body

- observing limitations on how many hours they can use the X-ray machines.

Ultraviolet light

Ultraviolet (UV) light is emitted from the Sun. Exposure to UV light from the Sun can give us a suntan.

However, too much exposure to UV light can cause damage to the skin, or increase the risk of cancer. UV light also causes the skin to produce vitamin D, which is important for the development of healthy bones. Sun cream should be applied regularly to absorb the UV light and protect the skin.

Table 12.4 *Frequency and wavelength range of ultraviolet light*

Ultraviolet light	
Frequency range	Between 7.5×10^{14} and 3×10^{16} Hz
Wavelength range	Between 400 and 10 nm

Note: $1\,nm = 10^{-9}\,m$

⟳ Keep up to date!

In August 2016 the Scottish Government issued health advice to give guidance on how to ensure sufficient intake of vitamin D. Search online to find out how much exposure to the Sun is recommended by the Scottish Government to help increase vitamin D production.

❓ Did you know ...?

X-rays were discovered almost by accident by German physicist Wilhelm Röntgen in 1895. He did not know what they were so called them X-rays ('x' is usually used for the unknown quantity in a maths problem). He received the Nobel prize in 1901 for his discovery.

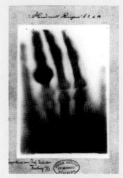

Figure 12.5 *One of the first X-ray photographs, taken by Wilhelm Röntgen of his wife's hand. The large black blob on the second finger is her wedding ring.*

📖 Word bank

- **Radiographer**
Radiographers are specially trained medical workers who operate gamma and X-ray machines used in medical imaging.

⁑ Make the link

There is information on how to convert between metres and other prefixes in the Investigation Skills Appendix.

☀Physics in action: Making the invisible visible

UV light is invisible to the eye, but fluorescent materials can absorb the energy in UV light and re-emit it as visible light.

Fluorescent lighting is often found in homes, schools and offices, either in compact fluorescent lights (CFL) or in strip lighting. Fluorescent lighting uses tubes filled with a gas which emits UV light when switched on. The UV is absorbed by a white phosphor coating on the inside of the lamp. This phosphor coating then re-emits the absorbed energy as white light.

Figure 12.6 *Ceiling lights in offices and schools often use fluorescent tubes.*

Some laundry detergent manufacturers advertise products which make clothes appear "whiter than white". Additives are added to the washing powders that absorb the invisible UV light from the sun and then fluoresce, emitting a bluish light. This makes the clothes appear whiter in sunlight. This is why some clothes appear to glow under artificial UV lighting.

The properties of UV light can be used in security features on credit cards, banknotes and passports to reduce forgery. Specially printed markings contain chemicals that fluoresce when illuminated with UV light. These markings are often hard for forgers to reproduce, so forgeries which do not have the marking are easy to detect.

UV marker pens produce ink that **fluoresces** under UV light but appear transparent in normal visible light. They are used to mark valuable property in homes and offices for security. If items are stolen, the ownership of the property can be identified. Police Scotland advise using UV pens to mark your bike so it can be identified if it is stolen.

Figure 12.7 *An official driver's licence has many security markings only visible under UV light.*

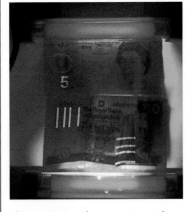

Figure 12.8 *The security markings on bank notes prove they are not counterfeit when viewed under UV light.*

📖 Word bank

- **Fluoresce**

  A fluorescent object is one that absorbs light and then re-emits it, usually at a longer wavelength. This process is known as fluorescence and the object re-emitting light is said to fluoresce.

Activity 12.1

 1. Experiment In this activity, you will investigate the use of UV markers and security markings.

Read the plan carefully then follow the instructions, making sure to follow all of the safety rules as you would normally be expected to do when carrying out practical work.

You can work in small groups but you should record your own observations and complete the answers yourself.

You will need: a UVA lamp, a UV marker pen, some paper, access to bank notes, credit cards, a driver's licence or a passport. You may also have access to UV beads.

Method

(a) Working with a partner, each write a message on a piece of paper with a UV marker pen. Swap messages. Then using the UV lamp, check if you can read each other's message.

(b) Collect various bank notes from different banks, credit or debit cards, passports or a driver's licence.

A teacher may be able to provide you with a driver's licence or access to different bank cards.

Use the UV lamp to examine each card or bank note and determine what markings are imprinted on it.

A secret message written with a UV marker pen, only visible under UV light.

(c) Explain how these markings are only visible under UV light and why they are included on these items.

Visible light

Visible light makes up just a small part of the electromagnetic spectrum. It is the only part of the EM spectrum that can be detected by the eye; it is this property that allows us to see the world. Visible light can also be detected by photographic film and digital light sensors. Visible light is not hazardous, although our eyes can be damaged if we look directly at the Sun.

Table 12.5 *Frequency and wavelength range for visible light*

Visible light	
Frequency range	Between 4×10^{14} and 7.5×10^{14} Hz
Wavelength range	Between 750 and 400 nm

Note: 1 nm = 10^{-9} m

⟳ Keep up to date!

Laser pointers produce intense beams of light that can be damaging to our eyes. In February 2016, over 30 laser pointer attacks were carried out on planes above Glasgow. The British Airline Pilots Association called for lasers to be classed as an offensive weapon. Search online to find out why laser pointers are dangerous to our eyes, and the precautions taken by governments to restrict their use. Try to find out what useful applications lasers have had since their invention in 1960.

Figure 12.9 *Barcode scanners use lasers to read the label on items in shops.*

Infrared

All hot objects emit infrared (IR) radiation. Although IR radiation cannot be seen by our eyes, infrared cameras use light sensors to detect infrared. The pictures they produce (called thermograms) show hot areas in red and cold areas in blue. Infrared radiation can also be detected by thermometers.

Table 12.6 *Frequency and wavelength range of infrared light*

Infrared light	
Frequency range	Between 3×10^{11} and 4×10^{14} Hz
Wavelength range	Between 1 mm and 750 nm

Note: 1 nm = 10^{-9} m

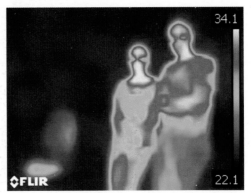

Figure 12.10 *A thermogram of the authors, taken with an infrared camera. Notice how the chair on the left is still warm, having been used moments before the photo was taken.*

Advances in thermal imaging in recent years have enabled medical workers to use imaging for diagnosis and treatment. For example, increased skin temperature can indicate that the tissue under the skin is inflamed.

Figure 12.11 *Infrared radiation penetrates into the tissue to increase blood flow.*

One application of infrared radiation in medical treatment uses the heat generated by infrared lamps to stimulate increased blood flow in a specific area, which can help relieve muscle pain.

Infrared radiation emitted from the body can be used to trigger burglar alarm systems and security lights. TV remote controls use coded pulses of infrared radiation to transmit information to the TV.

Figure 12.12 *TV remotes send data via short pulses of IR radiation. Each button has a different pattern that the receiver is programmed to recognise.*

> ### Make the link
>
> The heat from the Sun is transferred to the Earth using infrared radiation. Check Chapter 8 for more information on heat transfer by infrared radiation.

GO! Activity 12.2

1. **Experiment** In this activity, you will use the sensor on a smartphone camera to detect infrared radiation, and investigate the limitations of IR remotes.

Read the plan carefully then follow the instructions, making sure to follow all of the safety rules as you would normally be expected to do when carrying out practical work.

You can work in small groups but you should record your own observations and complete the answers yourself.

You will need: a smartphone/tablet camera, TV or music player remote control and a TV/ music player.

Method

(a) In a darkened room, open the camera app on the phone or tablet. Point the TV remote control toward the camera lens and press a button on the remote. What do you see on the screen of the camera?

> ### 🔍 Hint
>
> Does it travel through your hand, a sheet of paper, this textbook, or your smartphone?

(b) Using the TV/music player and its remote, investigate what materials the IR can travel through to activate the TV/music player.

(c) Plan and carry out an experiment to investigate the distance that the signal from the remote can be from the TV and still operate it.

2. Based on your investigations, describe some of the limitations of using IR remote controls. Do the limitations outweigh the advantages of an IR remote control?

> ### 🔍 Hint
>
> Include in your plan a procedure to find out if the IR signal can travel around walls.

Figure 12.13 *Satellites transmit and receive microwaves to transfer TV and data signals.*

Figure 12.14 *Tablets and smartphones have aerials to send and receive microwaves for Wi-Fi, Bluetooth and other forms of data connection.*

Microwaves

Microwave radiation is lower in frequency than infrared radiation. One of the most common applications of microwaves is found in the kitchen. Microwave ovens use microwaves to vibrate water and fat molecules in food and in drinks to heat them up.

Table 12.7 *Frequency and wavelength range of microwaves*

Microwaves	
Frequency range	Between 3×10^8 and 3×10^{11} Hz
Wavelength range	Between 0.1 cm and 100 cm

Note: $1 \text{ cm} = 10^{-2} \text{ m}$

Microwaves are also used for communication. They can be focused into narrow beams and used to transmit data between satellites orbiting the Earth and ground stations. Microwaves are detected with a parabolic dish aerial.

Modern data transfer technologies such as Bluetooth and Wi-Fi use microwaves.

 Activity 12.3

 1. **Experiment** In this activity, you will use a microwave transmitter and receiver to investigate microwaves.

Read the plan carefully then follow the instructions, making sure to follow all of the safety rules as you would normally be expected to do when carrying out practical work.

You can work in small groups but you should record your own observations and complete the answers yourself.

You will need: a microwave transmitter, a microwave receiver and a metal sheet for reflecting microwaves.

Method

Your teacher will set up and control the microwave transmitter.

(a) Set up the transmitter and receiver opposite one another, about 50 cm apart. When the transmitter is turned on, check to see if microwaves are detected at the receiver.

(b) Move the receiver to the left, and then to the right. What happens to the strength of the microwave signal when the receiver is not aligned with the transmitter?

Microwave transmitter directed toward microwave receiver.

(c) Set up the metal sheet to act like a mirror. When the transmitter is directed toward the metal sheet, move the receiver to a position where reflections might be expected. If the receiver picks up microwave signals, what does this suggest about microwaves?

Microwave transmitter and receiver directed toward metal reflector.

Make the link

Microwave signals obey the law of reflection, described in Chapter 11.

(d) Remove the metal sheet. What happens to the signal received by the microwave receiver?

Radio waves

Radio waves have the lowest frequency of all EM waves, and have the lowest energy. Their low frequency means they have very long wavelengths.

Table 12.8 *Frequency and wavelength range of radio waves*

Radio waves	
Frequency range	Between 3000 and 3×10^{11} Hz
Wavelength range	Between 0.1 cm and 100 000 m

Radio waves are used for TV and radio communication. Table 12.9 shows the different wavelengths that are used for different types of communication.

Table 12.9 *Wavelengths used for different types of communication*

Type of communication	Wavelength
Satellite TV	a few tens of centimetres
FM radio	a few metres
AM radio	tens of metres

Long-wave (AM) radio stations use very long wavelength radio waves that can bend around hills. This means that long-wave radio signals can travel very long distances.

Radio waves are detected with an aerial.

? Did you know ...?

Radio reception in the Highlands is often poor because of the mountains, which block the radio waves from transmitters. For example, the radio signal can often be lost when travelling along the A9 from Perth to Inverness and Wick. Switching from short-wave FM radio stations to long-wave stations increases the chance of keeping a signal.

GO! Activity 12.4

😊😊 **1. Experiment** In this activity, you will investigate the range of radio waves using walkie-talkies. You will also determine how radio waves can be blocked.

Read the plan carefully then follow the instructions, making sure to follow all of the safety rules as you would normally be expected to do when carrying out practical work.

You can work in pairs but you should record your own observations and complete the answers yourself.

You will need: two walkie-talkies, a tape measure and a metal biscuit tin.

Method

(a) Do this part of the experiment outside (find a large open space such as a car park, sports field or playground). Starting next to your partner, walk apart talking to each other on the walkie-talkie. Does the signal travel the whole length of the space you are in?

(b) Do this part of the experiment inside. Repeat part (a) in a long corridor. Does the signal travel as far as it did outside?

(c) Try to communicate with each other from different parts of the school building. Investigate different areas – do the radio waves travel around corners, or through walls?

(d) In the classroom, place one receiver in a tin box, and speak into the other. Does the walkie-talkie receive the signal? Put one receiver in a plastic box. Does the walkie-talkie receive the signal in the plastic box?

Walkie-talkies use radio waves to communicate.

2. Compare the advantages of using radio waves to communicate with the limitations discovered in this investigation.

> 🔍 **Hint**
>
> You will know if the signal is received if it makes a sound in the box.

GO! Activity 12.5

😊 **1.** Copy and complete the following table showing the regions of the electromagnetic spectrum. The regions increase in frequency from left to right.

Radio		Infrared	Visible		X-rays	

2. State two things that all regions of the electromagnetic spectrum have in common.

3. On hot summer days, the UV light in the Sun's rays can cause sunburn. State a method of reducing the risk of harm due to the Sun's UV light.

N3 L3 N4 L4

4. Copy the table with the following headings. Complete the table by stating an application, a hazard associated with exposure, and a method of detection for each type of EM radiation.

EM radiation	Application	Hazard	Detection
Gamma rays			
X-rays			
Ultraviolet			
Visible light			
Infrared			
Microwaves			
Radio waves			

5. For two of the types of EM radiation listed in the table for Q4:
   (a) describe one way that it has positively impacted our society
   (b) describe one way that it has improved our quality of life.

Learning checklist

After reading this chapter and completing the activities, I can:

N3 L3 N4 L4

- describe limitations associated with each type of electromagnetic radiation beyond the visible. **Activity 12.2 Q4, Activity 12.3, Activity 12.4** ○ ○ ○

- identify the different parts of the electromagnetic spectrum as gamma rays, X-rays, ultraviolet, visible, infrared, microwaves and radio waves. **Activity 12.5 Q1** ○ ○ ○

- state approaches to minimising risk for hazardous regions of the electromagnetic spectrum. **Activity 12.5 Q3** ○ ○ ○

- give examples of applications associated with each type of electromagnetic radiation. **Activity 12.5 Q4** ○ ○ ○

- state hazards associated with each type of electromagnetic radiation. **Activity 12.5 Q4** ○ ○ ○

- identify how the invisible parts of the electromagnetic spectrum can be detected. **Activity 12.5 Q4** ○ ○ ○

- describe how electromagnetic radiation has impacted upon society and our quality of life. **Activity 12.5 Q5** ○ ○ ○

13 Nuclear radiation

This chapter includes coverage of:

N4 Nuclear radiation • Forces, electricity and waves SCN 3-11b • Planet Earth SCN 04-04a • Materials SCN 4-15a• Topical science SCN 4-20b

You should already know:

• that nuclear energy is a non-renewable energy source.

National 4

Curriculum level 4

Materials: Properties and uses of substances SCN 4-15a

Atoms and radiation

Learning intentions

In this section you will:

• describe a simple model of an atom
• state that nuclear radiation can be absorbed by the material through which it travels
• investigate sources of background radiation, both natural and man-made
• describe safety procedures necessary when handling and storing radioactive substances.

📖 Word bank

• **Protons, neutrons, electrons and the nucleus**

Protons are positively-charged particles in the nucleus of an atom.

Neutrons are particles with no charge in the nucleus of an atom.

Electrons are negatively-charged particles which move around the nucleus of an atom.

Protons and neutrons are bound together to form the **nucleus**. The plural of nucleus is **nuclei**.

All the matter in the Universe is made up of **atoms**. An atom consists of a positively-charged centre called the **nucleus**. This is made of positively-charged **protons** and neutral **neutrons**. The nucleus is surrounded by moving, negatively-charged **electrons**.

Atoms are tiny. The diameter of an atom is about 10^{-10} m, or about one millionth of the width of a human hair. The nucleus is even smaller. The nucleus of an atom is about 10^{-15} m in size. This is about 1/100 000th of the size of the whole atom. The atom is mostly empty space. However, the neutrons and protons which make up the nucleus are much heavier than the moving electrons, so most of the mass of the atom is located in the nucleus.

A substance that is made of only one type of atom is an **element**. The properties of elements are determined by the different numbers of protons, neutrons and electrons that make up the atoms. The elements are organised according to their properties in the Periodic Table.

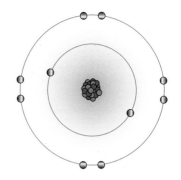

Figure 13.1 *The structure of an atom with protons and neutrons in the nucleus surrounded by orbiting electrons. (Not drawn to scale.)*

Radioactive decay and types of radiation

The nuclei in most elements are stable. Sometimes the nucleus is unstable, leading to **nuclear** or **radioactive decay**. Nuclei can decay by emitting different types of radiation:

- **alpha** radiation
- **beta** radiation
- **gamma** radiation.

The different types of radiation have different properties:

- they have varying energies
- they can be absorbed over different distances
- they can pass through different materials.

Their different properties mean that they can be used in a range of different applications and need different precautions to be taken when working with them.

> ### 📖 Word bank
>
> - **Alpha, beta and gamma**
> Alpha, beta and gamma radiation are all types of nuclear radiation.

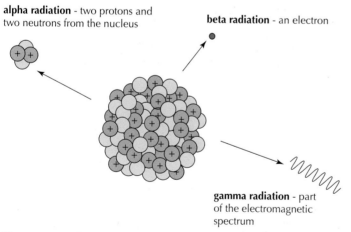

alpha radiation - two protons and two neutrons from the nucleus

beta radiation - an electron

gamma radiation - part of the electromagnetic spectrum

Figure 13.2 *Three types of radiation can be emitted from an unstable nucleus.*

Nuclear radiation is the energy that is transferred as radioactive materials decay. We are exposed to some radiation all the time. Radiation has a wide range of uses. It can be used in medicine or to produce electricity. It can also have a devastating effect when used in nuclear weapons.

Figure 13.3 *The mushroom cloud of the atomic bomb that exploded over Nagasaki, Japan during the Second World War.*

Absorption of radiation

When an atom **decays**, energy is emitted from the nucleus. The energy is emitted as alpha, beta or gamma radiation. The energy is absorbed as it passes through materials. The amount of energy absorbed depends on:

- the type of radiation
- the thickness of the material
- the type of absorbing material.

Table 13.1 *Properties of the three types of nuclear radiation.*

Name	What it is	How far it travels in air	Material it can be absorbed by	Speed of travel	Ionising effect
Alpha	Two protons and two neutrons from the nucleus	Around 8 cm	Paper	Slow (15×10^6 m/s)	Strong
Beta	An electron emitted from the nucleus due to the decay process	Around 1 m	Thin sheet of aluminium	Fast (270×10^6 m/s)	Weak
Gamma	A wave of energy, part of the electromagnetic spectrum	Very far (can be up to hundreds of metres in air)	Several centimetres of lead, concrete	The speed of light (300×10^6 m/s)	Very weak

❓ Did you know …?

When radiation passes through matter it can remove electrons from atoms in the matter by a process called **ionisation**. Alpha radiation is the most ionising radiation. Alpha radiation does not travel very far and can be absorbed by the material in a thin sheet of paper.

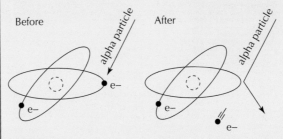

Figure 13.4 *Ionisation by an alpha particle.*

Beta radiation does not travel very far and is absorbed in thin sheets of metal, such as aluminium, and by sheets of glass or plastic. Beta radiation travels at about 90% of the speed of light and only has a weak ionising effect.

Gamma radiation can travel many kilometres through the air and can pass straight through the body, causing some ionisation damage. If ionisation happens to atoms inside cells inside the body, it can damage the DNA, leading to mutations.

⚛ Make the link

Chapter 12 has information about the frequency and wavelength of gamma radiation.

GO! Activity 13.1

☻ 1. Copy and complete the paragraph describing the structure of an atom, using the words provided.

elements   protons   Neutrons   atom   negative   electrons   positive

An _ _ _ _ has a central nucleus, which contains _ _ _ _ _ _ _ and neutrons. Protons have a _ _ _ _ _ _ _ _ charge. _ _ _ _ _ _ _ _ have neutral charge. Around the nucleus there are orbiting _ _ _ _ _ _ _ _ _. These have a _ _ _ _ _ _ _ _ charge. The properties of different _ _ _ _ _ _ _ _ depend on the different numbers of protons, neutrons and electrons that make up the atoms.

2. A pupil is examining the effect of different materials that could be used to shield the amount of radiation received from a radioactive source.

Material in front of radioactive source	Amount of radiation received (count rate)
Air	973
Sheet of paper	971
2 cm aluminium	444
10 cm lead	8

(a) Which material absorbed the most radiation?

(b) Which type of radiation (alpha, beta or gamma) do you think was used in the experiment? Explain your answer.

> 🔍 **Hint**
>
> Look back at Table 13.1. Which radiation is not stopped by air, paper or aluminium?

Background radiation

We are exposed to nuclear radiation around us all the time. This is known as **background radiation**. The average annual dose of radiation to someone living in Scotland in 2010 was 2300 microsieverts. This level of radiation does not harm humans.

> **?** **Did you know ...?**
>
> Radiation can be measured in units of **Sieverts**. This is the unit for the measurement of health impacts of ionising radiation, sometimes expressed in terms of **equivalent dose**. This takes into account the amount of radiation absorbed and the medical effects of the radiation.

> 📖 **Word bank**
>
> • **Background radiation**
>
> Background radiation is the radiation we are all exposed to all the time. It comes from both natural and man-made sources.

> **?** **Did you know ...?**
>
> The level of background radiation in Aberdeen is roughly double that in the rest of Scotland. This is because of all the granite used in buildings. Granite is a rock which produces the radioactive gas **radon**. Radon is only present in low levels and is not generally a cause for concern. Parts of Cornwall and Yorkshire also have higher levels of background radiation than the rest of the UK.

Figure 13.5 *The city of Aberdeen.*

GO! Activity 13.2

☻ Search online to find out the most radioactive places on Earth. Write a short report. Try to find out levels of background radiation where you live and compare them with the most radioactive places.

📖 Word bank

- **Natural radiation**

Natural radiation is the background radiation from natural sources.

- **Artificial radiation**

Artificial, or man-made, radiation is the background radiation from man-made sources.

Background radiation comes from a variety of sources. Around 86% of the radiation a person receives each year comes from **natural radiation**. In addition, **artificial** or **man-made radiation** from sources such as medical treatments and waste, and from nuclear power and nuclear weapons tests, contributes to background radiation.

The radiation dose received by an individual increases if they undergo medical treatment that involves radiation, such as radiotherapy used for treating cancer.

Table 13.2 *Examples of natural and man-made background sources.*

Natural sources of background radiation	Man-made sources of background radiation
Radon gas in the atmosphere	Medical treatments (radiotherapy, X-rays)
Cosmic rays from space	Nuclear power and weapon tests
The food and drink we eat	Low-level industrial and medical waste
Rocks and buildings	

❓ Did you know ...?

Most people are surprised to learn that some of the food they eat is radioactive. As plants grow they take up nutrients from the soil. The nutrients are stored within the plant and are transferred to our bodies when the plants are eaten. Some plants store higher levels of elements that can be radioactive. For example, bananas contain potassium, which is very slightly radioactive. However, it has been calculated you would need to eat 10 million bananas at once to receive a dangerous dose of radiation!

The radiation received from one banana is the same as one microsievert. Having a CT scan in hospital delivers the same dose as 70 000 bananas.

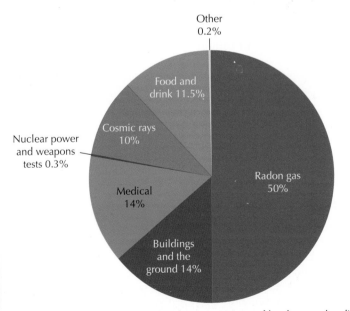

Figure 13.6 *Pie chart showing the proportions of background radiation sources.*

GO! Activity 13.3

☺ **1. Experiment** In this activity you will use
☺☺ a Geiger–Müller tube to measure
background radiation.

A Geiger–Müller tube (sometimes called a Geiger
counter) can be used to detect radiation.
Radiation entering the detecting tube causes
ionisation in the gas in the tube. The electrons
produced by the ionising event are accelerated to
the anode inside the tube. These ionising events
are counted and a measurement of the amount of
radiation detected is displayed.

*Measuring background radiation using a Geiger
counter.*

Read the plan carefully then follow the
instructions, making sure you follow all of the
safety rules as you would normally be expected
to do when carrying out practical work.

Record your own observations.

You will need: a Geiger–Müller tube.

Method

  (a) Switch on the Geiger–Müller tube. Measure and record the number of counts of
  radiation recorded in one minute.

  (b) Reset the counter and repeat the experiment 10 times.

  (c) Record your results in a table similar to the one below.

Test number	Number of counts per minute
1	
2	
3	
4	
5	
6	
7	
8	
9	
10	
Average	

  (d) Calculate the average background radiation.

  (e) Repeat the activity in a different location in the school. Is there a difference in the
  background radiation?

Nuclear safety

Nuclear radiation can damage or kill cells in living things. Radioactive materials must be handled and stored with care. The radioactive warning sign must be displayed where radioactive sources are used regularly or where they are stored. People working with radioactive materials should limit their exposure to radiation and follow strict safety procedures:

- never touch a source with bare hands – use tongs or special gloves to handle materials

- point the source away from the body

- never point the source at eyes

- wash hands after using radioactive sources

- use shielding, such as lead aprons

- wear film badges

- increase the distance to the source to receive less radiation.

Figure 13.7 *The radiation warning sign.*

☼ Physics in action: Film badges

A film badge is used to measure the levels of radiation to which the person wearing the badge is exposed. Film badges use photographic film. This film turns white when exposed to radiation. In a film badge, the film is cased behind different materials which absorb different types and levels of radiation. After the film has been used, it is developed and analysed to show the type and level of radiation to which the badge has been exposed. It is very important to follow safety procedures when working in scientific occupations, to control the risks and the hazards identified.

Figure 13.8 *A radiographer with a film badge.*

Radioactive waste

Radioactive material has to be stored carefully, even after it has been used. The material used in a nuclear power station is radioactive for a long time after it has been used to generate electricity.

Uranium-235 used in nuclear reactors is radioactive for over 700 million years. After it has been used as a fuel in the reactor, the material is stored in cooling ponds for about 20 years. This absorbs some of the radiation. It is then sealed in containers and stored in protected vaults. This is expensive and there are concerns about radiation leaks.

Most radioactive waste comes from nuclear power stations, but there are other sources of radioactive waste:

- medical waste (from diagnostic tools such as X-rays and tracers, and radioactive treatments such as cancer radiotherapy)

- industrial waste (from industrial X-rays, gauges and tracers)

- scientific research.

Figure 13.9   *Radioactive waste.*

> **GO!   Activity 13.4**
>
> ☺ **1.** Draw a simple model of an atom and label the protons, neutrons and electrons.
> **2.** Give an example of natural radiation.
> **3.** Give an example of man-made radiation.
> **4.** List three safety procedures needed when handling radioactive sources.
> **5.** Why does some radioactive waste have to be stored for a long time?

Uses of radiation

Curriculum level 3
Forces, electricity and waves: Vibrations and waves SCN 3-11b
National 4
Curriculum level 4
Forces, electricity and waves: Vibrations and waves SCN 4-11b

> **Learning intentions**
>
> In this section you will:
> - explore the different applications of nuclear radiation used in medicine
> - explore the different applications of nuclear radiation used in industry.

Nuclear radiation has many uses. It is used in medical applications to diagnose problems and as a treatment, to damage or kill harmful cells. It is used in industry for processes such as thickness monitoring, quality control and sterilisation.

Medical uses

Nuclear radiation is used for different types of medical processes:

- diagnosis – finding health problems (using e.g. radioactive **tracers** and X-rays)

- treatment – improving health problems (such as treating cancer).

Nuclear radiation can be used as a tracer to study the flow of liquids around the body, in the gut or in the blood stream. For example, if a doctor suspects there is a blockage in a kidney, such as a kidney stone, a radioactive source in a liquid can be

> **📖 Word bank**
>
> • **Tracer**
> Radioactive material can be used as a tracer to detect blockages. The radioactive material is injected inside the body and a gamma camera is used to detect the radiation on the outside.

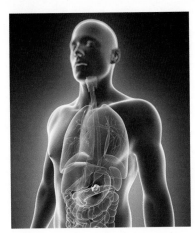

Figure 13.10 *Nuclear radiation used as a tracer.*

injected. This can be detected from outside the body using a gamma camera. A blockage will cause the radioactive source to collect at the site of the blockage. This will give a larger reading at that point. This technique is called **tracing.** Tracing can also be used to search for cancerous tumours.

X-rays are used to take images of the inside of the body. X-rays are a form of electromagnetic radiation, and they have similar properties to gamma rays. Bones and different types of tissue in the body absorb different amounts of radiation, so photographic images taken with X-rays will show these differences. Doctors and radiographers use the X-ray images to help with their diagnosis of complaints such as broken bones, tooth decay, pneumonia and lung cancer.

Cancer cells are cells in the body that have mutated and grown out of control. There are more than 100 types of cancer affecting different parts of the body. Cancer cells must be destroyed to prevent the cancer from spreading further.

Radiotherapy is commonly used to treat cancer. Radiotherapy machines aim beams of gamma rays at the cells, which are destroyed by the energy of the gamma rays. Treatment has to be carefully planned to prevent excessive doses of radioactivity. The beam of gamma rays is moved through different angles so that healthy tissue around damaged cells is not damaged by receiving too much radiation.

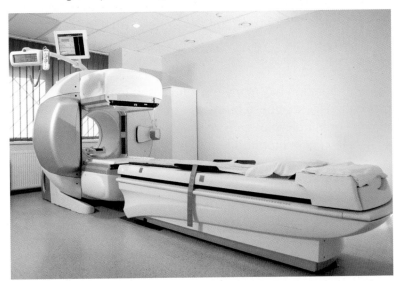

Figure 13.11 *Equipment used in radiotherapy. The bed remains stationary while the head of the machine rotates.*

The source of radiation used in medical procedures must be carefully selected. It must emit gamma radiation:

- with sufficient energy so that it will pass through the body

- that will decay quickly so that the patient is not exposed for longer than necessary.

Industrial uses

Industrial uses of radiation involve the detection of radiation using:

- Geiger–Müller tubes
- scintillation counters.

Leak detection

Geiger–Müller tubes are used to detect leaks in buried pipes. A radioactive tracer is added to the liquid or gas in the pipe. If there is a crack in the pipe, the liquid or gas will leak into the surrounding soil and the radiation can be detected using a Geiger–Müller tube. This shows the position of the crack in the pipe without having to dig up the whole pipe.

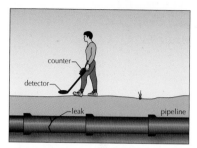

Figure 13.12 *A radioactive tracer in a leaking pipe can be detected above ground. The leak can be found without having to dig up the whole length of the pipe.*

Thickness monitoring

The thickness of material, such as aluminium foil, can also be checked using nuclear radiation. This process uses the absorption of radiation as it passes through solid materials. A radioactive source is placed above the material as it is being processed (in this case, being rolled to a required thickness). A Geiger–Müller tube under the material records the intensity of the radiation that passes through the material. If the intensity is too high (too much radiation getting through), the material is too thin. If the reading is too low (not enough radiation getting through), the material is too thick. The machines in the manufacturing process can be altered to correct the mistake.

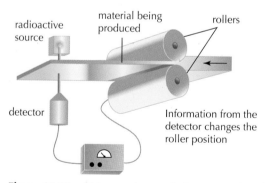

Figure 13.13 *Using nuclear radiation to monitor the thickness of a material.*

Cargo monitoring

Scintillation counters count the number of light signals arising from the ionising effect of radiation. Scintillation counters are used in freight terminals and border security posts to monitor cargo, to check for nuclear material being transported illegally. Scintillation counters are also used in radiation survey meters for environmental monitoring.

Smoke alarms

Figure 13.14 *A smoke alarm.*

Some smoke alarms have a small source of radiation inside the plastic shell. This source causes ionisation in the air, which allows a small current to flow in a circuit. Smoke particles reduce the current in the circuit by absorbing alpha radiation, and the electronics cause the alarm to sound. The amounts of radiation used in smoke detectors are very small compared with normal background radiation.

GO! Activity 13.5

☻ Search online to find out more information about medical and industrial applications of nuclear radiation.

GO! Activity 13.6

☻ 1. Describe one use of nuclear radiation in medicine.
   2. Describe one use of nuclear radiation in industry.

National 4

Curriculum level 4

Planet Earth: Energy sources and sustainability SCN 4-04a;
Topical Science SCN 4-20b

Nuclear energy

Learning intentions

In this section you will:

• discuss advantages and disadvantages of generating electricity using nuclear fuel
• compare the risk of nuclear radiation with other environmental hazards.

Make the link

Nuclear power stations are explained in more detail in Chapter 1.

Much of the electricity used in Britain is generated in large thermal power stations. Thermal power stations all work in the same way. A fuel source is burned to heat water and form steam, which is then used to drive turbines in an electricity generator. Some power stations use fossil fuels (coal and gas) as the raw fuel. Other power stations use radioactive materials as the heat source in a nuclear reactor.

In most nuclear reactors, uranium is used as the fuel. Most uranium is mined in Kazakhstan, Canada, Australia, Niger, Namibia and Russia. Uranium ore is processed to produce uranium dioxide fuel rods. These fuel rods release large amounts of energy through radioactive decay in nuclear reactors. This energy heats water to create steam, which drives the turbines.

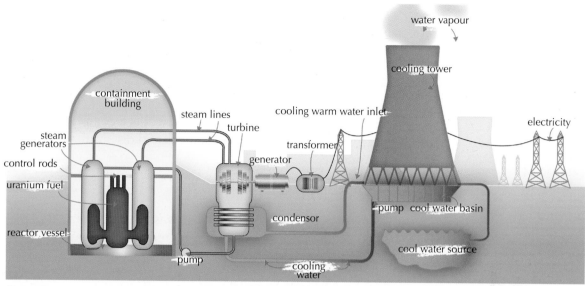

Figure 13.15 *A diagram of a nuclear power station.*

Advantages and disadvantages of nuclear generation

There are advantages and disadvantages to using nuclear material to generate electricity. Table 13.3 lists some of the important features.

Table 13.3 *Some of the advantages and disadvantages to nuclear power stations.*

Advantages	Disadvantages
Does not use fossil fuels, which are running out	There is a finite amount of nuclear fuel – it will eventually run out
Does not produce carbon dioxide, so does not contribute to global warming	Nuclear waste is difficult and expensive to dispose of and store safely
1 kg of nuclear fuel produces about 3 million times as much energy as 1 kg of coal (Only uses a small amount of fuel to produce electricity)	Nuclear waste remains radioactive for thousands of years
	Nuclear disasters can be very serious (Chernobyl, Fukushima)

Most nuclear power stations in Europe were built in the 1970s and 1980s. Some countries, such as France, use nuclear power stations for a major part of their electricity generation. Other countries have decided to stop using nuclear power. For example, Germany is phasing out nuclear power and plans to close all its nuclear power plants by 2020. They are decommissioning nuclear power stations and investing money in alternative methods of electricity generation.

In 2008, the Scottish Parliament voted against the construction of any new nuclear power stations in Scotland. There are two nuclear power stations in Scotland, one at Torness in East Lothian and one at Hunterston, in North Ayrshire. In 2015, these two power stations generated about 35% of the electricity generated in Scotland.

Choosing whether to use nuclear power is a complicated decision with conflicting advantages and disadvantages to be considered.

Nuclear waste

Nuclear energy generates nuclear waste. Used (spent) fuel is the most obvious type of waste. There are three levels of nuclear waste:

- low-level waste

- intermediate-level waste

- high-level waste.

Table 13.4 *Types of waste from nuclear power stations.*

Type of radioactive waste	Examples
Low-level	Contaminated materials and equipment such as paper, tools, industrial clothing, air filters.
Intermediate-level	Chemical sludges, metal cladding and contaminated materials.
High-level	Used fuel rods and chemicals from reprocessing fuels.

Low-level waste is put in drums and encased in concrete. The drums are then buried in landfill sites. Some low-level waste can be incinerated to reduce its volume.

Intermediate-level radioactive waste is considered hazardous and has to be stored and monitored carefully.

High-level waste remains radioactive for thousands of years. It is stored in water in large pools for 20 years, and is then stored and guarded in purpose-built underground stores. High-level waste decays into intermediate-level waste over many thousands of years.

Nuclear waste has to be transported from power stations to storage facilities. Many people are concerned about the risk of accidents happening while the waste is being transported to storage. The environment is at risk from events and accidents such as radiation contamination, fuel leaks or safety procedures not being followed.

| N3 | L3 | N4 | L4 |

Fossil fuel or nuclear generation?

All electricity generation has advantages and disadvantages. Power stations which use fossil fuels as the energy source cause environmental damage because they emit greenhouse gases which contribute to global warming. By contrast, nuclear generation does not produce greenhouse gases, but it can cause other types of environmental damage in the event of accidental leakage of radiation, and the problems of long-term storage of high-level waste have not yet been fully solved.

? Did you know ...?

Even when there are strict safety procedures, nuclear accidents can happen. In March 2011, a huge tsunami hit the Fukushima nuclear power plant in Japan. The cooling systems in the power plant were seriously damaged. Without the cooling systems the reactors overheated. This caused a serious radiation leak. Around 160 000 people within a 30 km radius from the power station were evacuated from their homes and there is still a no-go area around the site.

Figure 13.16   *The explosion and nuclear radiation leak from the Fukushima power plant, Japan.*

GO! Activity 13.7

Search online to find out about the Chernobyl accident in 1986. What impact did this accident have on farmers in Britain?

GO! Activity 13.8

Plan and hold a debate on the use of nuclear power. Should radioactive material be used to generate electricity? Search online to find out more about this issue. Think about safety, cost, environmental issues and the demand for cheap electricity. If you have a strong opinion for or against, can you see the argument from someone else's point of view?

Learning checklist

After reading this chapter and completing the activities, I can:

N3 L3 N4 L4

- state that an atom has a central nucleus containing protons and neutrons, and orbiting electrons. **Activity 13.1 Q1, Activity 13.4 Q1** ○ ○ ○

- explain that nuclear radiation can be absorbed by the material through which it travels. **Activity 13.1 Q2** ○ ○ ○

- state that background radiation is all around us and comes from natural and man-made sources. **Activity 13.3** ○ ○ ○

- give examples of the different ways that radioactive substances have to be handled and stored. **Activity 13.4 Q4, Q5** ○ ○ ○

- give examples of applications of nuclear radiation used in medicine. **Activity 13.5, Activity 13.6** ○ ○ ○

- give examples of applications of nuclear radiation used in industry. **Activity 13.5, Activity 13.6** ○ ○ ○

- state that nuclear radiation can be used in power stations to generate electricity and give some of the advantages and disadvantages of this method. **Activity 13.7, Activity 13.8** ○ ○ ○

- explain that the risks of nuclear power stations have to be compared to other environmental hazards. **Activity 13.7, Activity 13.8** ○ ○ ○

Unit 2 practice assessment

National 3 Outcomes

N3 Wave properties

1. Complete the following paragraph.

    Waves transfer _____. The highest point of a wave is called the
    _____. The _____ point of a wave is called the trough. The
    _____ is the distance between _____ identical points of a wave.
    The amplitude is _____ the total height of the wave.   **6**

N3 Light

2. Using the phrase 'light travels in straight lines', explain how a shadow is
    formed when light hits an object.   **2**

3. The following diagrams show rays of light being affected by different
    optical instruments. Match each diagram to the optical instrument that
    would create that effect.   **6**

    triangular prism   **plane mirror**   **convex lens**

    periscope   **convex reflector**   **rectangular glass block**

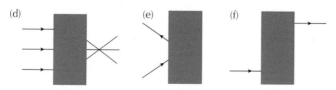

4. The signage on the front of this fire engine states
    the words FIRE - RESCUE. The Fire Brigade wish to
    replace the sign with the words laterally inverted.

    (a) Write how the signage should look when the
    words are printed laterally inverted.   **1**

    (b) Why is it helpful for other drivers to have the
    sign written in this way?   **1**

5. The lens in the eye focuses the light on the retina at the back of the eye to create an image.

   (a) The lens refracts the light. Explain what this means. **1**

   (b) The eye in this figure is not focusing the light correctly on the retina.

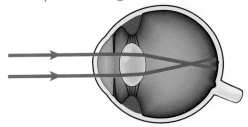

      (i) State the name of the sight defect. **1**

      (ii) Explain what type of lens would correct this sight defect. **1**

   (c) A camera has many similarities to the eye. Describe two things a camera and the eye have in common. **2**

N3 Colour

6. Three coloured spotlights are shone onto a small area so they overlap. Copy and complete the diagram to show the colours that are formed when the red, green and blue spotlights overlap. Label the colours produced. **2**

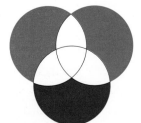

7. When apple trees come into blossom in the spring, they have green leaves and blossoms which are pink when they are closed and white when they open.

   (a) What colours of white light are reflected from:

      (i)   the leaves? **1**

      (ii) the closed pink blossom? **1**

      (iii) the opened flower petals? **1**

   (b) Bees collect pollen from the yellow part of the flower, called the stamen. Why does the stamen appear yellow? **1**

N3 Optical instruments

8. The following optical instruments use either reflection or refraction to function: digital cameras, periscopes, telescopes, microscopes, binoculars.

   (a) Select an optical instrument from the list above that uses reflection to function and explain how it works. **1**

   (b) Select an optical instrument from the list above that uses refraction to function and explain how it works. **1**

N3 Electromagnetic radiation

9. Ultraviolet waves are part of the electromagnetic spectrum. State a benefit to the human body gained by exposure to ultraviolet waves.   **1**

10. The table shows details of the radio frequencies used for different radio stations.

Station	FM waveband (short wave)	MW waveband (medium wave)	LW waveband (long wave)
Radio Scotland	92–95 MHz	810 kHz	not available
Radio Wales	93–102 MHz	657 kHz	not available
Radio Ulster	92–95 MHz	1341 kHz	not available
Radio 4	103–105 MHz	1449 kHz	198 kHz

(a) A lorry driver tunes his radio to Radio Scotland.

   (i) State a possible radio wave frequency that the radio is receiving.   **1**

   (ii) He takes the ferry from Cairnryan in Scotland to Belfast, in Northern Ireland. Although he didn't retune his radio, he now finds he is listening to Radio Ulster. Suggest a reason for this.   **1**

(b) A different lorry is travelling to Inverness from Glasgow, and travels through the Highlands. The driver is originally listening to Radio 4 on the FM waveband, but loses reception. What other waveband would you recommend the driver tune into so they can continue listening to Radio 4? Give a reason for your answer.   **2**

11. State an application of X-rays and describe how X-rays have had a benefit on society.   **2**

National 4 Outcomes

N4 Wave characteristics

1. What is meant by the term 'amplitude' and how could it be measured?   **2**

2. Name a type of energy that is transferred by a transverse wave.   **1**

3. Name a type of energy that is transferred by a longitudinal wave.   **1**

4. A frequency of 200 Hz is another way of saying 200 waves per second. True or false?   **1**

5. If the transverse water wave shown in the diagram is moving from left to right, what direction will the particles move?   **1**

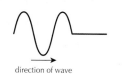
direction of wave

6. A swimming pool has a total length of 50 m and there are 16 complete waves along the length of the pool. Calculate the wavelength of one wave.   **3**

7. A wave has a frequency of 300 Hz and a wavelength of 0.5 m. Calculate the speed of the wave.   **3**

8. A wave travels 2 m in 0.08 seconds. Calculate the speed of the wave. **3**

9. The wave shown travels 4.0 m in 7 seconds. Calculate the:

   (a) amplitude **1**

   (b) wavelength **3**

   (c) frequency **3**

   (d) speed. **3**

N4 Sound

10. A piano produces a note that has 256 vibrations per second.

   (a) Are these waves transverse or longitudinal? **1**

   (b) Can you determine the volume of the sound from the information given? **1**

11. What will happen to the amplitude of a waveform if a sound is played louder? **1**

12. A guitar plays two notes and the sounds are shown on a screen:

   (a) Which of the images shows the higher frequency sound? **1**

   (b) The sound from the guitar is heard 5 s after it was produced. If sound travels at 340 m/s, calculate the distance travelled by the sound. **3**

13. Describe the method for measuring the speed of sound using the echo from a wall. State the measurements that need to be taken and how to determine the speed from the results. **3**

14. What piece of apparatus is needed to measure sound level? **1**

15. Why should sound level be monitored in factories? **1**

16. A student makes the following statements about hearing protection:

   A No sound is too loud

   B Reducing the volume of the sound can protect your hearing.

   C Reducing the time listening to the sound can protect your hearing.

   Indicate which of these statements are correct **2**

17. State two applications of ultrasound. **2**

N4 Electromagnetic spectrum

18. Different types of waves in the electromagnetic spectrum are used in medicine and industry.

   (a) State one way in which microwaves are used in the communication industry. **1**

   (b) State one use of infrared radiation in medicine. **1**

   (c) State a hazard associated with ultraviolet radiation. **1**

   (d) State a precaution a radiographer might take when operating an X-ray machine. **1**

19. Read the following passage.

   In a doctor's surgery, infrared ear thermometers have replaced traditional oral thermometers for measuring body temperatures of small children and babies. Infrared ear thermometer measure the infrared heat generated by the eardrum, which accurately reflects the core body temperature. The infrared thermometer can take accurate readings in less than one second. They also work when the child or baby is asleep.

   Oral thermometers are slower, require a certain technique to work, and often disturb sleeping children. In the past oral thermometers contained mercury, which is poisonous.

   Another type of IR thermometer called a temporal thermometer can be used to measure body temperature. It detects the infrared radiation from the body by pointing the thermometer at the forehead. However, temporal thermometers can sometimes provide inconsistent readings, and pointing them in the wrong direction can give inaccurate readings.

   (a) Using information given in the passage state **two** advantages of the infrared ear thermometers for measuring body temperature in babies. **2**

   (b) Temporal thermometers also detect infrared radiation from the body. State a reason why using an ear thermometer is better than using a temporal thermometer for measuring body temperature. **2**

N4 Nuclear radiation

20. State a natural source of nuclear radiation. **1**

21. State a man-made source of nuclear radiation. **1**

22. Give an example of nuclear radiation used in medical applications. **1**

23. Give an example of nuclear radiation used in industrial applications. **1**

24. State one advantage of using nuclear fuel to generate electricity. **1**

25. State one disadvantage of using nuclear fuel to generate electricity. **1**

UNIT 3

Dynamics and space

14 Speed and acceleration

This chapter includes coverage of:

N3 Forces • N4 Speed and acceleration • Forces, electricity and waves SCN 4-07a • Forces, electricity and waves SCN 4-07b

You should already know:

- that objects travel at different speeds.

National 3

National 4

Speed, distance and time

Learning intentions

In this section you will:

- learn that speed is a measure of the distance covered by an object in a given time
- use the relationship between speed, distance and time
- learn that there is a difference between average and instantaneous speed.

Speed is a measure of the distance travelled by an object in a given time. The speed of a vehicle can be calculated by measuring the distance travelled, and dividing by the time taken.

$$\text{speed} = \frac{\text{distance}}{\text{time}}$$

Make the link

We use this equation to calculate speed in other chapters, for example in Chapter 10 (Sound) and Chapter 16 (Satellites).

Speed is commonly measured in these units:

- metres per second, m/s
- kilometres per hour, km/h
- miles per hour, mph.

In physics, the standard unit for speed is metres per second, or m/s. This can also be written as ms^{-1}. All the units include a distance part and a time part.

A car travelling at 60 mph is said to be travelling faster, or at a higher speed, than a car travelling at 30 mph. The car travelling at 60 mph will cover twice the distance of the car travelling at 30 mph in the same time.

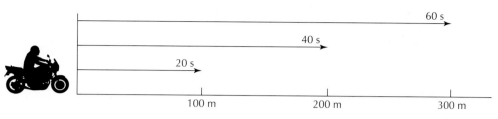

Figure 14.1 *A bike travelling at constant speed.*

Figure 14.1 shows a bike travelling at a constant speed:

- in the first 20 s, the bike travels 100 m – its speed is 5 m/s

- in the first 40 s, the bike travels 200 m – its speed is 5 m/s.

So the speed is constant for different parts of the bike ride.

For an object travelling at constant speed, every distance divided by the time taken to travel that far will give the same value. Here the bike is travelling at 5 m/s. Table 14.1 shows the speed of some different objects.

Table 14.1 *Speeds of different objects*

Object	Speed (m/s)
Snail	0.0028
Person walking	1.3
Usain Bolt running the 100 m at the Rio Olympics	10.2
Fast road car (Bugatti Veyron 16.4 Super Sport)	115
Boeing 747 aeroplane in flight	240

Think about a time you have been in a car or a bus going through a town. There are times when you are moving and times when you have stopped and are stationary, for example at traffic lights and at road junctions. The journey could go quickly or you could get stuck for hours.

The speed taken for the whole trip is the **average speed**. The speed at any moment during the journey is the **instantaneous speed**.

A car speedometer shows the speed of a car speed at a particular time. This is a measure of the instantaneous speed.

? Did you know ...?

Richard Noble is a Scottish entrepreneur who is developing the Bloodhound SSC. This is a high-speed land vehicle, designed to travel at speeds up to 1600 km/h. It is powered by a jet engine and a rocket.

Figure 14.2 *The Bloodhound SSC.*

📖 Word bank

- **Average speed**

Average speed is the speed measured over a large distance or a long time.

- **Instantaneous speed**

Instantaneous speed is the speed at any given instant in time.

Figure 14.3 *A car speedometer.*

Answers to all activity and assessment questions in this Unit are available online at www.leckieandleckie.co.uk/page/Resources

GO! **Activity 14.1**

1. Define the term 'speed'.
2. Which two quantities are needed to work out the speed something is travelling at?
3. Two motorbikes travel 70 miles. The first bike takes 2 hours to complete a journey, the second takes 3 hours.

   Which one has the higher speed?
4. Put these things in order, from slowest to fastest:
   a space shuttle, a snail, a car, a sprinter, an aircraft, a baby crawling.

Calculating speed

The average speed measures the total distance travelled divided by the total time. Speed is calculated using te equation shown here.

$$\text{speed} = \frac{\text{distance}}{\text{time}}$$

This can be written

$$v = \frac{d}{t}$$

Table 14.2 *Symbols and units used in the speed equation.*

Name	Symbol	Unit	Unit symbol
Average speed	v	metres per second	m/s
Distance	d	metres	m
Time	t	seconds	s

Example 14.1

A runner travels 100 metres in 10 seconds. Calculate his average speed.

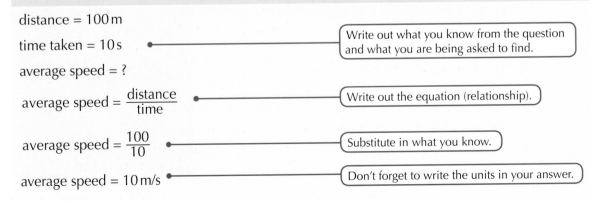

distance = 100 m

time taken = 10 s — Write out what you know from the question and what you are being asked to find.

average speed = ?

average speed = $\dfrac{\text{distance}}{\text{time}}$ — Write out the equation (relationship).

average speed = $\dfrac{100}{10}$ — Substitute in what you know.

average speed = 10 m/s — Don't forget to write the units in your answer.

N3 | L3 | N4 | L4

Example 14.2

A snail has an average speed of 0.013 m/s. Calculate how far it will travel in 60 seconds.

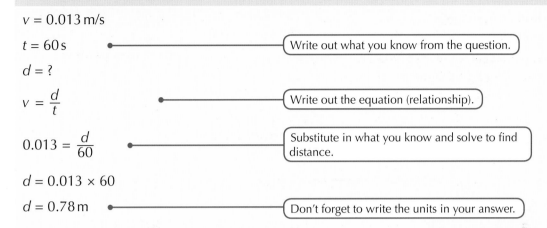

$v = 0.013$ m/s

$t = 60$ s — Write out what you know from the question.

$d = ?$

$v = \dfrac{d}{t}$ — Write out the equation (relationship).

$0.013 = \dfrac{d}{60}$ — Substitute in what you know and solve to find distance.

$d = 0.013 \times 60$

$d = 0.78$ m — Don't forget to write the units in your answer.

Make the link

See the Investigation Skills Appendix for more about writing equations using the symbols.

GO! Activity 14.2

😊 **1.** Using the relationship for average speed, complete the missing values **(a)–(g)** in the table.

Distance (m)	Time (s)	Speed (m/s)
25	5	**(a)**
40	10	**(b)**
(c)	12	144
5000	**(d)**	25
10 000	**(e)**	25
230	2	**(f)**
0.3	10	**(g)**

2. A car travels a distance of 20 metres in a time of 5 seconds. Calculate the average speed of the car.

3. A ball travels a distance of 18 m in a time of 4 s. Calculate the average speed of the ball.

4. A train travels from Dundee to London (700 km) in a time of 6 hours. Calculate the average speed of the train in km/h.

5. Richard cycles to work every day. His journey of 19 kilometres takes him 40 minutes. Calculate his average speed in metres per second. 1 km = 1000 m, 1 minute = 60 seconds

Activity 14.3

1. Experiment In this activity you will use a toy car on a sloping ramp to measure average speed. This will involve measuring distances and times.

Read the plan carefully then follow the instructions, making sure you follow all of the safety rules as you would normally be expected to do when carrying out practical work.

You should work in a small group but record your own observations and complete the answers yourself.

You will need: a ramp, a lab jack or pile of books, a toy car, a ruler and a stopwatch.

Method

(a) Set up a slope or ramp using a lab jack or a pile of books to give a gentle slope.

(b) Mark three points on the slope:
   - one close to the top, labelled **start**
   - one a little further down the slope, labelled **A**
   - one at the bottom of the slope, labelled **B**.

   Choose a distance between A and B which is easy to use in calculations. Measure and record the distance.

(c) Put a toy car at the top of the slope so it is level with your start line.

(d) Let the car roll down the slope – don't push it.

(e) Start the stopwatch when the front of the car passes line A.

(f) Stop the stopwatch when the front of the car passes line B and record the time in a table like the one below.

(g) Complete the table to show your results and calculate the speed.

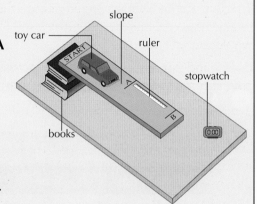

Slope number	Distance from A to B (cm)	Time to travel from A to B (s)	speed = $\dfrac{\text{distance}}{\text{time}}$
1			

(h) How could you change the speed of the car? What could you change about the apparatus?

(i) Pick one of your ideas from part **(h)** and repeat the experiment. You will need to extend your table to fit your new results.

(j) Is this method of measuring speed accurate? What could you do to improve your experiment?

Make the link

Thinking about what you will measure in an experiment is all part of the planning stage. Read through the Investigation Skills Appendix for more information on experiments.

Average and instantaneous speed

Learning intentions

In this section you will:

- use the relationship between speed, distance and time
- state that there is a difference between average and instantaneous speed
- investigate different ways of calculating average and instantaneous speed.

Remember that average speed of an object is calculated using this equation:

$$\text{average speed} = \frac{\text{distance}}{\text{time}}$$

To calculate the average speed of a journey, you would use the total distance of the whole journey and the total time for the whole journey.

During the journey, the speed will increase and decrease. At any moment in time, the object has an instantaneous speed.

Speed cameras

Speed cameras on roads can work out average speeds (over a long distance or time), or instantaneous speeds (over a short distance or time).

Average speed cameras (SPECS cameras) use special technology to read car number plates as each car passes different locations. The distance between the camera locations is known. The time taken to travel between any two locations is measured and the average speed is calculated. The cameras use infrared technology so that they can work in the dark as well as during daylight.

Instantaneous speeds are measured using LiDAR (Light Detection And Ranging) speed guns.

Police traffic officers use a LiDAR speed gun to measure the speed of a target vehicle. The LiDAR gun fires pulses of laser light at the surface of a vehicle. A sensor on the gun measures the time it takes for each pulse to bounce back. Using the fact that the pulses of laser light travel at a constant speed of 300 000 000 m/s, the distance to the vehicle can be calculated. This distance and the time taken by the light pulses are used to calculate the speed. This instrument works with short distances and so gives an instantaneous speed.

✹ Physics in action: Stopping distances

Drivers have to be aware of their speeds when driving on the road. They also have to be aware of their surroundings so that they can slow down and stop safely when needed.

In an emergency, the total distance travelled before stopping is the **stopping distance**.

The stopping distance has two components:

- thinking distance
- braking distance.

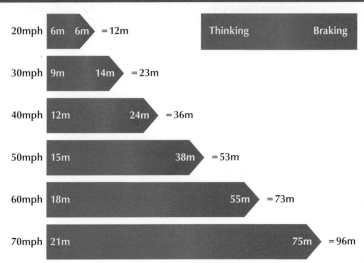

	Thinking	Braking
20mph	6m 6m	= 12m
30mph	9m 14m	= 23m
40mph	12m 24m	= 36m
50mph	15m 38m	= 53m
60mph	18m 55m	= 73m
70mph	21m 75m	= 96m

Figure 14.4 *Stopping distance for different speeds of car.*

The **thinking distance** is the distance the car travels in the time it takes for the driver to make the decision to press the brake pedal in order to stop.

The **braking distance** is the distance the car travels in the time it takes for the brakes to bring the car to a stop.

Both the thinking distance and the braking distance are affected by the speed of the car. The faster the car is going, the longer it takes to stop. This is an example of the speed relationship in action. If the tyres are good and the road is dry, the frictional forces acting on the car will slow it down.

Measuring average speed

Average speed can be found from a whole journey. It is measured over a long distance or over a long period of time.

✹ Make the link

See the Investigation Skills Appendix to learn about making experiments more accurate and reliable.

GO! Activity 14.4

☺ Plan an experiment to find the average speed of a person running the length of a playing field. Think about the apparatus you will need and the measurements you will have to take. How could you make your experiment accurate and reliable?

Measuring instantaneous speed

Measuring instantaneous speed is harder than measuring average speed because of the need to measure short periods of time or small distances. The instantaneous speed of a vehicle can be measured by finding the average speed during a very short time. The shorter the time, the closer it will be to an instant.

N3	L3	**N4**	L4

It is difficult to measure short times accurately with an ordinary stopwatch because human reaction time affects the measurements. To overcome this problem, electronic timers are used.

Electronic timers use a beam of infrared light between a transmitter and a receiver. When the beam is broken, the receiver sends a signal to the electronic timer and the timer starts. When the light is detected again, the receiver sends a second signal to stop the timer.

If the length of the object that cut the light beam is known, the speed can be calculated using the equation:

$$\text{speed} = \frac{\text{length of object}}{\text{time taken through beam}}$$

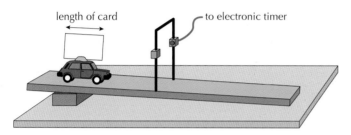

Figure 14.5 *Using an electronic timer to measure instantaneous speed.*

Light gates can be connected to computers. The user enters the length of the object into the computer program and the speed is calculated automatically and can be displayed instantly.

Sporting events use electronic timers to determine the time taken to finish a race. A laser beam is shone across the finish line to a receiver. The time is recorded when the laser beam is broken by a runner crossing the line. Timers used in sporting events can measure times to thousandths of a second.

GO! Activity 14.5

☻ Search online to investigate the use of electronic timers, light gates and speed cameras.

You might want to think about:

- what information can they provide?
- what research or development is being carried out using electronic timers?
- how can they benefit our lives?

GO! Activity 14.6

😊😊 **1. Experiment** In this activity you will use a toy car on a sloping ramp equipped with a light gate to measure instantaneous speed.

Read the plan carefully then follow the instructions, making sure you follow all of the safety rules as you would normally be expected to do when carrying out practical work.

You should work in a small group but record your own observations and complete the answers yourself.

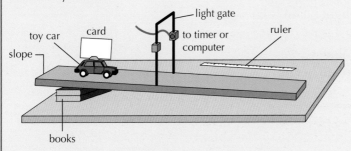

Using an electronic timer to measure instantaneous speed.

You will need: a ramp, a lab jack or pile of books, a toy car with a small piece of card stuck to the top, a ruler and a light gate timer connected to a computer.

Method

(a) Set up a slope or ramp using a lab jack or a pile of books to give a gentle slope.

(b) Place your light gate across the ramp so that the car can travel freely through it.

(c) Connect the light gate to a timer or a computer. This will measure the time the car takes to travel through the light gate.

(d) Prepare the piece of card to be stuck to the top of the car. Cut the card to a length which is easy to use in calculations.

(e) Attach the card to the top of the car. Make sure the card will pass through the light gate when it rolls down the slope.

(f) Let the car roll down the slope – don't push it – so the card goes through the light gate.

(g) Read and record the time shown on the timer or computer.

(h) Calculate the instantaneous speed of the card using the relationship for speed, distance and time.

(i) Repeat the experiment. Do you get the same result? Why might the results be different?

(j) Investigate what happens to the car's speed if you increase the angle of the ramp. Use a protractor to measure the angle of the slope. Draw a table to record your results.

Activity 14.7

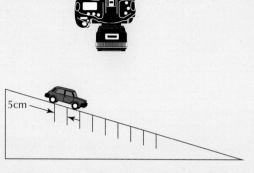

1. Experiment In this activity you will use a toy car on a sloping ramp equipped with a digital camera to measure instantaneous speed. You need a digital camera with a continuous drive mode so it can automatically take photographs at high speeds.

The continuous drive mode of a digital camera means you can use photographs of a moving object on a suitable background to work out how far the object travels between each photograph. If you know the time between the photographs, you can use the distance and the time to calculate the speed.

Photographing the motion of a moving object.

You might also be able to download an app for your phone that would allow you to use your smartphone camera. You would still need to know the time interval between the photos.

Read the plan carefully then follow the instructions, making sure you follow all of the safety rules as you would normally be expected to do when carrying out practical work.

You should work in a small group but record your own observations and complete the answers yourself.

You will need: a toy car or dynamics cart, a ramp, a digital camera, a support stand or tripod with a moveable head, tape or glue, and a large sheet of paper.

Method

(a) Mark lines on the sheet of paper at set distance apart (every 5 cm).

(b) Put the marked paper on the ramp. Fix it with tape or glue to make sure it doesn't move.

(c) Set your digital camera to take multiple photographs at high speed. Most cameras will have a setting of 10 fps (frames per second). This gives a time interval of 0.1 s.

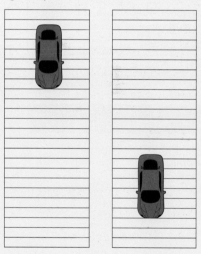

Here the car has crossed 13 lines on the lined paper between camera shots.

(d) Use the support stand or tripod to arrange the camera directly above the lined paper.

(e) Let the car roll down the slope. Use the camera shutter button to take repeated photos as the car goes down the slope. It is probably easiest if one person releases the car and another person operates the camera.

(f) Use the review function of the camera to compare two photographs. Work out the distance travelled between pictures.

(g) Use the equation speed = $\dfrac{\text{distance}}{\text{time}}$ to find the instantaneous speed of the car.

GO! **Activity 14.8**

☺ **1.** Explain why your average speed and your top speed over a car journey will be different.

2. What might be the benefit to road safety to have average speed cameras rather than instantaneous ones?

National 4
Curriculum level 4
Forces, electricity and waves: Forces SCN 4-07a

Speed–time graphs

Learning intentions

In this section you will:

- draw speed–time graphs to describe motion
- calculate the distance travelled from a speed–time graph.

📖 **Word bank**

- **Speed–time graph**

A speed-time graph shows the speed of an object at any given time.

The speed of an object over its journey can be represented on a **speed–time graph**. A speed–time graph shows the speed of an object at any given time. It is a useful way to describe the motion of an object. Time is always plotted along the x-axis and speed is plotted up the y-axis.

The shape of the graph shows whether the object is speeding up (accelerating), slowing down (decelerating) or moving at a constant speed. It can also show if the object is stationary.

⚬: **Make the link**

See the Investigation Skills Appendix to read about drawing graphs.

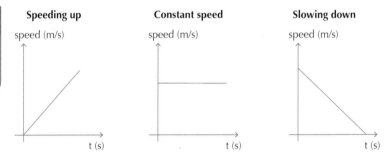

Figure 14.6 *Speed–time graphs.*

📖 **Word bank**

- **Acceleration**

Acceleration is the rate of change of speed. An object which is slowing down has negative acceleration, or deceleration.

The gradient (slope) of the line on a speed–time graph indicates the **acceleration**.

Example 14.3

The table shows data for a moving object or person.

Time (s)	0	2	4	6	8	10
Speed (m/s)	0	4	8	8	8	0

1. Plot a speed–time graph for this data.

2. What was the maximum speed?

3. Suggest what movement could be represented by the graph.

1.

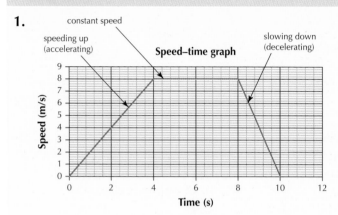

2. The maximum speed is 8 m/s.

3. The movement could be a person running a sprint:

- for the first 4 seconds, their speed increases from 0 m/s to 8 m/s

- from 4 s to 8 s, they are moving at a constant speed of 8 m/s

- from 8 s to 10 s, they are slowing down from 8 m/s to 0 m/s as they get to the end.

GO! Activity 14.9

☺ Plot a speed–time graph for each of these data sets, then answer the following questions.

1.

Time (s)	0	5	10	15	20	25	30
Speed (m/s)	2	4	6	8	10	12	14

2.

Time (s)	0	2	4	6	8	10	12
Speed (m/s)	0	5	10	15	20	25	30

3.

Time (s)	0	15	30	45	60	75	90
Speed (m/s)	6	6	6	6	6	6	6

(continued)

4.

Time (s)	0	10	20	30	40	50	60
Speed (m/s)	20	20	20	15	10	5	0

5.

Time (s)	0	3	6	9	12	15	18
Speed (m/s)	3.5	3.5	3.5	3.5	3.5	3.5	3.5

6. On each graph label any points where the object was stationary, moving at a constant speed or was speeding up and slowing down.

7. What was the maximum speed for each object plotted on the graphs you have drawn?

8. Which graph could represent:

   (a) driving away from traffic lights?

   (b) carefully reversing into a parking space?

   (c) travelling at a constant speed?

   (d) driving along a road then braking to avoid hitting a car?

   (e) accelerating to overtake a bicyclist?

Using speed–time graphs to find distances travelled

Speed–time graphs can be used to calculate the distance travelled by an object. The distance travelled is found by calculating the area under the line on the graph.

Example 14.4

Calculate the total distance travelled by the object represented in the speed-time graph

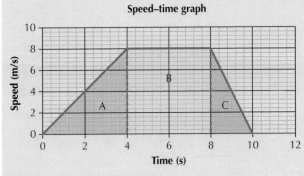

Speed–time graph

Area A = $\frac{1}{2}$ × base × height ●━━━━━━━━ (Calculate the area of each labelled area.)

$= \frac{1}{2} \times 4 \times 8$

Area A = 16

| N3 | L3 | **N4** | L4 |

Area B = base × height

= 4 × 8

Area B = 32

Area C = $\frac{1}{2}$ × base × height

Area C = $\frac{1}{2}$ × 2 × 8

Area C = 8

Total area = 16 + 32 + 8 = 56 •———— Find the total area by adding Area A + Area B + Area C.

The total distance travelled is 56 m. •———— Don't forget to write the units in your answer.

GO! Activity 14.10

☺ For each of the graphs below, calculate the total distance travelled by the object.

1.

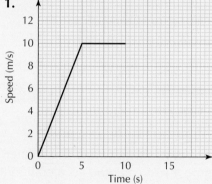

2.

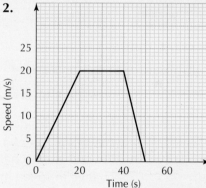

🔍 Hint

Divide each graph into simple shapes. Find the area of each shape and add to find the total.

3.

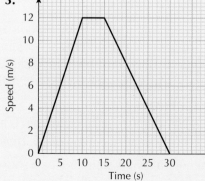

4.

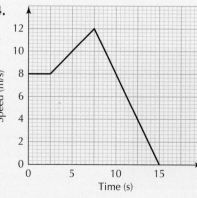

Speed–time graphs can be plotted using data-capturing devices connected to motion sensors. The motion sensors can take many readings and so plot very precise graphs. This gives a quick method for drawing graphs and analysing the speed of objects.

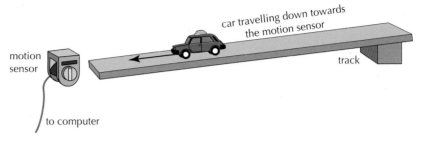

Figure 14.7 *Using a motion sensor to plot speed–time graphs.*

National 4

Curriculum level 4

Forces, electricity and waves: Forces SCN 4-07b

Acceleration

Learning intentions

In this section you will:

- learn that acceleration is the change in speed
- investigate how to measure acceleration
- use the relationship involving acceleration, change in speed and time.

Most vehicles do not travel at the same speed all the time. Acceleration describes how quickly speed changes. It is defined as the change in speed in a given time. The greater the change in speed in a set time, the greater the acceleration (or deceleration) of the object.

Acceleration is commonly measured in the units metres per second per second, or m/s².

$$\text{acceleration} = \frac{\text{change in speed}}{\text{time}}$$

The standard unit m/s² can also be written ms⁻².

The time taken for a car to accelerate from 0 to 60 mph is one way of comparing the performance of cars. Formula 1 racing cars have very high acceleration. They have been recorded to accelerate from 0 to 100 km/h in as little as 1.6 seconds. This gives an acceleration of 17.4 m/s².

The Tesla Model S electric car has a time to accelerate from 0 to 100 km/h of 2.8 seconds. This gives an acceleration of 10.3 m/s².

Speed–time graphs show acceleration. A steeper slope shows greater acceleration. A negative, or downwards, slope shows deceleration.

Figure 14.8 *Formula 1 race cars.*

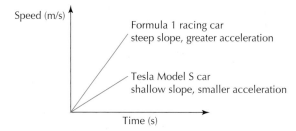

Figure 14.10 *Speed–time graph showing the acceleration of two different cars.*

Figure 14.9 *Tesla Model S electric car.*

Calculating acceleration

Acceleration can be calculated using the equation shown.

$$\text{acceleration} = \frac{\text{change in speed}}{\text{time taken for the speed to change}}$$

$$= \frac{\text{final speed} - \text{initial speed}}{\text{time taken for the speed to change}}$$

This can be written as

$$a = \frac{\Delta v}{t}$$

$$a : \frac{v - o}{t}$$

Δ is the Greek character 'delta', which is used to show a change in a quantity.

Table 14.3 *Symbols and units used in the acceleration relationship.*

Name	Symbol	Unit	Unit symbol
Acceleration	a	metres per second per second	m/s^2
Change in speed	Δv	metres per second	m/s
Time	t	seconds	s

Example 14.5

A car accelerates from rest to 9 m/s in 4 s. Calculate the acceleration.

initial speed = 0 m/s

final speed = 9 m/s •———— Write out what you know from the question.

time taken = 4 s

acceleration = ?

change in speed = final speed − initial speed •— Calculate the change in speed.

change in speed = 9 − 0 = 9 m/s

$$\text{acceleration} = \frac{\text{change in speed}}{\text{time taken for the speed to change}}$$ •— Write out the equation (relationship) for acceleration.

$$\text{acceleration} = \frac{9}{4}$$ •———— Substitute in what you know.

acceleration = 2.25 m/s² •———— Don't forget to write the units in your answer.

| N3 | L3 | **N4** | L4 |

Example 14.6

A van accelerates at 1.25 m/s² from 1 m/s to 4 m/s. Calculate the time taken.

initial speed = 1 m/s

final speed = 4 m/s ———————————(Write out what you know from the question.)

acceleration, a = 1.25 m/s²

time, t = ?

change in speed = final speed – initial speed ———(Calculate the change in speed, Δv.)

$\Delta v = 4 - 1 = 3$ m/s

$a = \dfrac{\Delta v}{t}$ ———————————(Write out the equation (relationship) for acceleration.)

$1.25 = \dfrac{3}{t}$ ———————————(Substitute in what you know.)

$t = \dfrac{3}{1.25} = 2.4$ s ———————(Don't forget to write the units in your answer.)

Measuring acceleration

To measure acceleration, we need to know:

- the initial speed

- the final speed of the object

- the period of time.

If an object is starting from rest, the initial speed is 0, so we only need to measure the final speed.

We can measure acceleration using light gates and electronic timers using one of two methods:

- Method 1 uses two light gates at different parts of a slope

- Method 2 uses a single light gate.

Method 1: two light gates

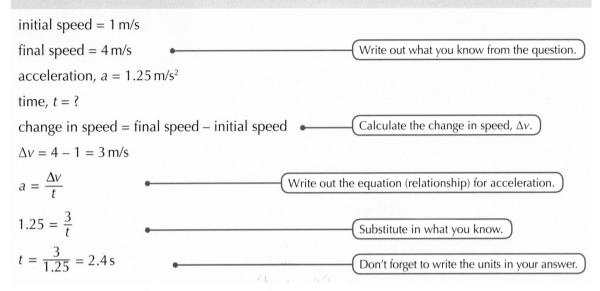

length of card · to electronic timer · stopwatch · light gate 1 · t_1 · light gate 2 · t_2

Figure 14.11   *Measuring acceleration using two light gates.*

The speed has to be measured at two different points on a slope. The light gates use the length of the card and the time taken to cut the light beam to find the speeds. The time taken to travel between the first and second light gate is also measured using a hand-held stopwatch.

This equation is used to calculate the acceleration:

$$\text{acceleration} = \frac{\text{speed through second light gate} - \text{speed through first light gate}}{\text{time taken to travel between the gates}}$$

Method 2: one light gate

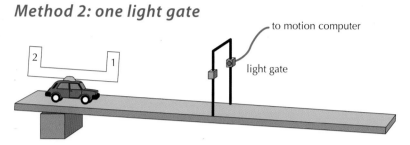

Figure 14.12 *Measuring acceleration using one light gate.*

Acceleration can also be measured using two sections of card and one light gate. The length of both sections of card must be measured. The light gate will find the time taken for both sections to cut the light beam. The light gate measures the time between the two cards going through the gate. The computer calculates the acceleration of the car using this equation:

$$\text{acceleration} = \frac{\text{speed of second piece of card} - \text{speed of first piece of card}}{\text{time between the two pieces going through the gates}}$$

GO! Activity 14.11

☻ **1.** What is meant by the term 'acceleration'?

2. What quantities have to be measured to calculate acceleration?

3. A car is travelling at 10 m/s when it accelerates for 2 s to a speed of 15 m/s. Calculate the acceleration.

4. A bus accelerates away from a set of traffic lights. It takes 5 seconds to reach 5 m/s. Calculate the acceleration.

5. A snail takes 40 seconds to move off from rest to reach a speed of 0.005 m/s. Calculate the acceleration.

6. A car starting from rest accelerates down a road at 1.75 m/s². It reaches a speed of 14 m/s. Calculate the time taken by the car to accelerate to that speed.

7. A train slows from 35 m/s to 20 m/s in 5 s. Calculate the deceleration. (Hint: find the change in speed and use the acceleration relationship.)

8. A car slows down from 16 m/s to 0 m/s in 8 s. Calculate the deceleration.

9. A bike is travelling at 6 m/s. It decelerates at 3 m/s² to come to a stop. Calculate the time taken to come to rest.

10. Search online to find out the acceleration values of different modes of transport.

GO! Activity 14.12

😊😊 If you have access to light gates and a suitable computer program, try measuring acceleration using either of the two methods described above.

1. Which method would give the least accurate measurement of acceleration? Explain you answer.

2. What would happen to the value of acceleration if you change the angle of the slope? Plan an experiment to test your hypothesis.

✦ Make the link

See Chapter 15 for more about the relationship between acceleration and force and how cars can be designed to be safer.

❓ Did you know ...?

When there is a car crash, the car, its contents and the passengers decelerate rapidly. Rapid acceleration (and deceleration) requires larger forces. The forces acting on the body during deceleration can be fatal. Pilots and rollercoaster riders also experience high forces during changes of direction and speed.

Figure 14.13 *Scottish racing driver Allan McNish crashed during the first hour of the 24 hour Le Mans race in 2011. He quickly decelerated from 140 mph and was unharmed.*

Learning checklist

After reading this chapter and completing the activities, I can:

N3 L3 N4 L4

	• state that speed is a measure of the distance covered by an object over a given time. **Activity 14.1 Q1**	○ ○ ○
• use the relationship between speed, distance and time. **Activity 14.2, Activity 14.3, Activity 14.4**	○ ○ ○
• investigate how to measure average and instantaneous speed. **Activity 14.3, Activity 14.6, Activity 14.7**	○ ○ ○
• draw speed–time graphs to describe motion. **Activity 14.9**	○ ○ ○
• calculate the distance travelled from a speed–time graph. **Activity 14.10**	○ ○ ○
• state that acceleration is the change in speed per second (or rate of change). **Activity 14.12 Q1**	○ ○ ○
• use the relationship involving acceleration, change in speed and time. **Activity 14.11 Q3–Q10**	○ ○ ○
• investigate how to measure acceleration. **Activity 14.12**	○ ○ ○

15 Forces, motion and energy

This chapter includes coverage of:

N3 Forces • N4 The relationship between forces, motion and energy • Forces, electricity and waves SCN 3-07a • Forces, electricity and waves SCN 3-08a • Forces, electricity and waves SCN 4-07b

You should already know:

- how friction affects the motion of moving objects
- how air resistance affects the motion of moving objects
- how to reduce friction to improve the efficiency of moving objects
- that objects can be moved by magnetic, electrostatic and gravitational forces.

Forces

National 3

Learning intentions

In this section you will:

- describe how forces are measured with a force meter
- learn that forces change the shape, speed or the direction of moving objects
- investigate the effect and application of forces.

Forces are all around us but we cannot see them. When forces cause movement it is possible to see their effects. However, even when something is not moving, there are still forces at work. A number of forces can be acting at the same time. Arrows can be used to show the size and direction of a force.

🔵GO! Activity 15.1

A book sitting on a table does not move. Are there any forces acting on it? What kind of forces could they be? If you pushed the book, now what forces are acting? Discuss your ideas in your group before writing a short summary of your thoughts.

Word bank

• Contact forces

Contact forces have to touch an object to have an effect on it.

• Non-contact forces

Non-contact forces have an effect on an object without touching it.

Forces can push, pull and turn. Some forces need to touch the object. These are called **contact forces**. To push a book across a table, you have to touch it.

Other types of forces can act over a larger distance. These are called **non-contact forces**. Gravity is a non-contact force which acts on every object.

Forces can be described depending on the effect they have on objects, such as:

- lift – the force that aircraft experience as a result of the movement of the wings through air. Lift acts upwards, in the opposite direction to the gravitational force.

- buoyancy – the force exerted on an object in a fluid (gas or liquid). Buoyancy also acts upwards, in the opposite direction to the gravitational force.

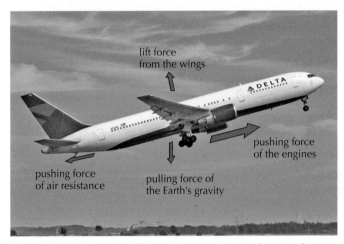

Figure 15.1   *There are different forces acting on the aeroplane as it takes off.*

Some non-contact forces work in areas of space known as **fields**. There are three types of fields we experience:

Make the link

Chapter 3 has more information on electromagnetism.

- gravitational fields

- magnetic fields

- electrostatic fields.

A mass in a gravitational field will experience a force of attraction. The Moon is in the Earth's gravitational field, and so the Moon and Earth are attracted to each other.

A magnetic material such as iron in a magnetic field will experience a force. Magnets have a magnetic field around them and can be used to pick up objects.

A charged particle in an electrostatic field will experience an attractive or repulsive force. Attractive forces pull things together, repulsive forces push them apart.

Figure 15.2 *The effect of a gravitational field.*

Figure 15.3 *The effect of an electrostatic field.*

Figure 15.4 *The effect of a magnetic field.*

Measuring forces

The unit of measurement for forces is the **newton**. The symbol for newtons is N. 10 N is 10 newtons.

Force can be measured using a **force meter**. This is also known as a newton meter or a newton balance. Force meters can be made in different sizes to measure small and large forces, and to measure forces with different degrees of precision.

> ### 📖 Word bank
>
> • **Force meter**
> A force meter measures forces.

> ### ❓ Did you know ...?
>
> Force meters use the extension of a spring to measure force. The behaviour of a spring when it is stretched is known as Hooke's Law. This law states that a spring will extend regularly as the force on it is increased.
>
>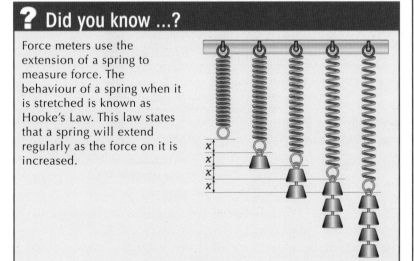
>
> **Figure 15.5** *The extension of a spring.*

> ### ❓ Did you know ...?
>
> Sir Isaac Newton was a famous British scientist who lived in the seventeenth century. He described three fundamental laws of motion. These laws are still applied in many areas including aeronautical engineering, space travel, transport safety and motor sports.
>
>
>
> **Figure 15.6** *Sir Isaac Newton (1643–1727).*

(continued)

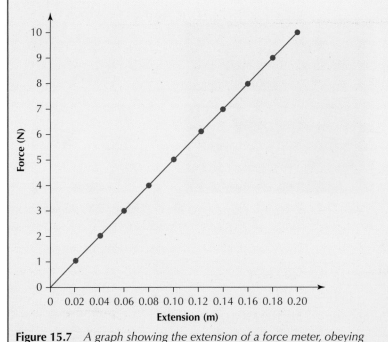

Figure 15.7 *A graph showing the extension of a force meter, obeying Hooke's law.*

Figure 15.8 *Different force meters.*

GO! Activity 15.2

😊😊 **1. Experiment** In this activity you will make and test your own force meter.

Read the plan carefully then follow the instructions, making sure you follow all of the safety rules as you would normally be expected to do when carrying out practical work.

You should work in a small group but record your own observations and complete the answers yourself.

You will need: a clamp stand, a clamp and bar, a spring, a ruler, 1 N weights on a hanger and a pointer to attach to the spring.

Method

(a) Measure the original length of an unstretched spring.

(b) Set up the apparatus as shown the diagram.

(c) Add 1 N to the spring and measure the new length.

(d) Calculate the extension of the spring (how much the spring has stretched when compared to the original length).

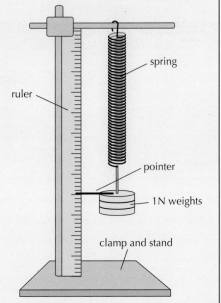

Experiment to test your own force meter.

(e) Continue adding 1 N to the spring until 5 N has been added, and find the extension each time. Use a table like the one below to record your results.

Force added (N)	Extension (cm)
1	
2	
3	
4	
5	

> **Make the link**
>
> See the Investigation Skills Appendix for more on drawing graphs with dependent and independent variables.

(f) Draw a graph of your results. Plot the independent variable, force, on the horizontal axis, and the dependent variable, extension, on the vertical axis.

(g) What conclusion can you make based on your results?

(h) Suggest a value for the extension of the spring if you added 6 N. Test you answer using the apparatus.

(i) What did you have to keep the same during your experiment? What was your control variable?

(j) What would happen if you continued to add a force to a spring? Would it keep on stretching?

2. Search online to find out what is meant by the term 'elastic limit'.

GO! Activity 15.3

☺ **1.** Copy and complete the following paragraph about forces, using the words provided.

newtons   **push**   **pull**   **friction**   **contact**   **force meter**   **turn**

Forces can _ _ _ _, _ _ _ _ and _ _ _ _. There are contact forces (e.g. _ _ _ _ _ _ _ _) and non- _ _ _ _ _ _ _ forces (e.g. gravity). Forces are measured with a _ _ _ _ _ _ _ _ _ _. Forces are measured in _ _ _ _ _ _ _.

2. A spring being tested stretches 2 cm when a force of 10 N is applied to it. If it stretches evenly, how far would you expect it to stretch if you added:

(a) 20 N?   **(b)** 2 N?

> **Hint**
>
> You could draw a graph to help you answer this question. Use the fact that the extension is 0 when the force is 0 to help plot the graph.

Shape, speed and direction

A force can change the shape, speed and direction of a moving object. For example, when a tennis racket makes contact with a ball it can:

- change the shape of the ball (it becomes squashed for a short period of time) and bend the racket strings

- change the speed of the ball (the ball will speed up or slow down)

- change the direction the ball is travelling in.

Activity 15.4

1. **Experiment** Plan an experiment to investigate the effects of forces. Do all forces change the shape, speed and direction of objects? You could:

   - make a small boat or trolley with a sail. Use a desk fan to investigate the size of force needed to make it move.

   - investigate the force needed to change the direction of a light paper ball. (Hint: search online to research the game 'paper toss'.)

   - investigate the force needed to change the shape of different-sized balls of plasticine.

   - investigate the thickness of material through which a magnet will apply a force to a paper clip.

Think carefully about the apparatus you will need and the measurements you will have to take. Can you measure the speed or acceleration of the object?

> **Make the link**
>
> See Chapter 14 for more about measuring speed and acceleration.

When you have made your plan, carry out the experiment. Make sure you follow all of the safety rules as you would normally be expected to do when carrying out practical work. You should work in a small group. Record your observations in the form of a poster. What did you do, what did you find out? Don't forget to evaluate your experiment.

Activity 15.5

1. What three things can forces do to objects?
2. What are the units of force?
3. Name a piece of apparatus used to measure forces.
4. Complete the table by predicting the size of force for the situation.

0.01 N		500 N		1 500 000 N
How much an ant can lift	To push a book across the desk	To lift your own weight	To move a car	To move a train

National 4

Measuring forces to keep us safe

Learning intention

In this section you will:

- learn how measurements of forces are used to keep us safe.

It is important to measure forces. Bridges, buildings and tunnels are all designed and tested to be able to carry the forces they may be subjected to when they are used. Cars, aeroplanes and trains all have to be tested to ensure that the forces during a possible impact will not be passed to the passengers.

Engineering safety

Bridges have to be designed and tested to withstand the forces of the vehicles passing over them. In December 2015, the Forth Road Bridge was closed for over a month after a crack was found in some of the support structures. It was essential to carry out repairs to prevent further damage, which could have resulted in a collapse of parts of the bridge.

Figure 15.9 *The Forth Road Bridge.*

 Activity 15.6

😊😊 **1. Experiment** In this activity you will measure the force a paper bridge can withstand before it collapses.

Read the plan carefully then follow the instructions, making sure you follow all of the safety rules as you would normally be expected to do when carrying out practical work.

You should work in a small group but record your own observations and complete the answers yourself.

You will need: five sheets of newspaper, sticky tape, a paper cup, some string or thread, a paper clip and some coins.

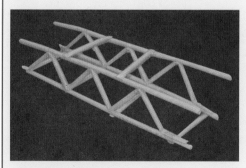

A paper-straw bridge.

Paper cup hanging from a paper-straw bridge.

Method

(a) Using the newspaper and sticky tape, create a structure that bridges a 1m gap between two benches. Think about the best way to use the paper – will it be strongest if you roll it into tubes, or fold it or layer it?

(b) Attach the cup to the bridge using the string and paper clip.

(c) Add coins to the cup. Keep adding coins until the structure of the bridge fails.

(d) What safety precautions did you have to consider when doing this experiment?

(e) If there was more than one design of bridge, which one withstood the biggest force? Can you explain why?

Figure 15.10 *Protective surface under playground equipment.*

Playground safety

Playgrounds are designed with protective surfaces to reduce the severity of injuries from falls. The protective surface absorbs the impact and reduces injury.

There are three types of protective surfacing for play areas: grass; loose material (such as bark); synthetic materials (such as rubber tiles).

GO! Activity 15.7

1. **Experiment** In this activity you will investigate the properties of different surfaces that could be used in play areas.

Read the plan carefully then follow the instructions, making sure you follow all of the safety rules as you would normally be expected to do when carrying out practical work.

You should work in a small group but record your own observations and complete the answers yourself.

You will need: a clamp and stand, a ruler, a marble and different surfaces such as foam, corrugated cardboard, a tray of sand, wood and grass.

Method

(a) Set up the apparatus as shown in the diagram.

(b) Drop the ball from a set height. Measure and record the height it bounces back to (its rebound height).

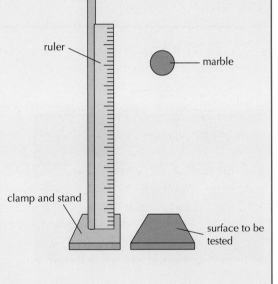

(c) Repeat your measurements before changing the surface the ball is dropped on. You could try a wooden floor, carpet, grass, concrete and bark. If possible, go to a local park and find the rebound height from the protective surface there.

(d) What kind of graph should you draw for these results? Draw a graph using your results.

(e) From these results, which surface would be best as the protective surface for a playground?

(f) How does dropping a ball onto a surface simulate playground activity?

(g) Why is it good to repeat the experiment and find the average?

> **Hint**
>
> To calculate an average add all the test results together and then divide by the number of tests you performed.

Car safety

Cars are designed to reduce the forces transmitted to the bodies of drivers and passengers in the event of a collision. Crash tests are used to show the effects of collisions and to experiment with different ways of reducing the impact.

Car designs use features such as airbags, roll cages, collapsible steering columns and crumple zones. These all change the forces experienced by the occupants of the car during impact, reducing potential injuries. By increasing the time it takes an object to stop, we can reduce the force of impact.

Figure 15.11   *A crash between two cars.*

Balanced forces

Learning intentions

In this section you will:

- explain how objects can travel at constant speed when balanced forces are acting upon them
- describe the term `friction'
- investigate the force of friction
- give examples of situations where friction is wanted and not wanted.

If two forces that are acting on an object are equal in size but in opposite directions, the forces cancel each other out. They are **balanced forces**. When two teams in a tug-of-war pull with the same-sized force and in opposite directions, the forces are balanced and the teams do not move. If one team pulls with a greater force, the forces are no longer balanced and the teams move.

If there are balanced forces acting on a moving object, it will carry on moving at the same speed.

A car travelling at a constant speed will have the drive force of the engine (the red arrow in Figure 15.13) balanced by the total friction acting on the car (the friction due to the tyres on the road and the air resistance; the green arrows in Figure 15.13).

The speed of the car will change when the forces are unbalanced:

- to increase the speed, the forward driving force has to be bigger than the opposing forces of friction and air resistance

- to decrease the speed, the forces of friction and air resistance have to be bigger than the driving force.

| National 3 |
| Curriculum level 3 |
| Forces, electricity and waves: Forces SCN 3-07a |
| National 4 |

📖 Word bank

- **Balanced forces**

The forces on an object are balanced if they have the same size and act in opposite directions. They cause objects to remain stationary or continue with a constant speed.

Figure 15.12   *Balanced forces acting on a box.*

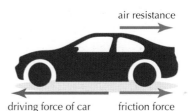

Figure 15.13   *Balanced forces on a car.*

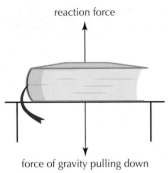

reaction force

force of gravity pulling down

Figure 15.14 *The forces acting on a book, resting on a table.*

Even when a book is sitting on a table, there are forces acting on it. The downward force of gravity is balanced by the reaction force given by the table, so the book does not move. The upwards and downwards forces are equal in size and opposite in direction. Newton summarised the idea of balanced forces in his First Law.

Newton's First Law

An object will remain at rest or continue moving with constant force when acted on by balanced forces.

Physics in action: There's no friction in space!

It is hard to find moving objects that have **no** forces acting on them on Earth as there are usually external forces. Our atmosphere provides a force of friction, called **air resistance**, as objects move.

Space rockets fly beyond the atmosphere, so they are not affected by air resistance. A space rocket uses rocket motors to accelerate to a high speed. When the motors are switched off, there are no forces acting on the rocket and it continues to move at the same speed. If rocket motors are used to change the position of the rocket as shown in Figure 15.15, the rocket will keep spinning until another force is applied.

Figure 15.15 *A space rocket using its motors to start spinning.*

Physics in action: Obeying the laws of physics in a car

You must wear a seatbelt when you are travelling in a car – it is the law! A seatbelt is an application of Newton's First Law. Consider a car suddenly stopping as it hits a wall. The driver's seat will stop suddenly because it is fixed to the car. However, the driver will keep moving at a constant speed unless there is something to apply a force to them. A seatbelt applies a restraining/opposing force to stop the driver moving and helps prevent injuries and fatalities.

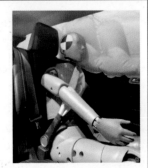

Figure 15.16 *Seatbelt on a crash test dummy.*

N3 L3 N4 L4

Friction

If you push a ball along the floor it will move for a distance and then come to a stop. According to Newton's First Law, the ball should move at constant speed unless there is an external force acting upon it. As the ball does come to a stop, there must be an external force in action. This external force is **friction**. It means there are no longer balanced forces acting on the ball.

Friction opposes movement. When an object is stationary it will not move until the pulling or pushing force is big enough to overcome friction. Friction always acts in the opposite direction to the direction in which the object is moving. The force of friction can also be called `drag' or `air resistance' when objects are moving in fluids.

Without friction, many everyday activities would be very difficult:

- your feet would slide backwards when you walk or run

- car and bike brakes would not slow you down

- clothes would slip through pegs on a washing line.

Figure 15.17 *Friction in action.*

Friction is a contact force that exists when two surfaces touch. No surfaces are completely smooth, so when two objects touch each other the tiny bumps and ridges on one surface can rest in the hollows of the other. For the surfaces to slide across each other the bumps and ridges must ride up out of the hollows; this needs a force. The smoother the surface, the less force needed to make the surfaces slide across each other.

Friction can be increased by:

- making surfaces rougher – mountain bike tyres have a rough tread to help grip to slippy road surfaces.

- increasing the area of contact – a larger brake disc will exert a bigger force and slow a bike down faster.

Too much friction can cause problems. In a car engine, moving parts can get too hot if friction forces are too high. This heat can become so intense it can melt metal parts or cause them to expand, stopping them from moving. On a bike, the chain and gear wheels are subject to friction, making it harder to ride the bike.

❓ Did you know ...?

Maglev trains use magnetic forces to lift and propel vehicles. Maglev stands for magnetic levitation. Magnets are used to lift the train from the track. There are no surfaces in contact with each other, so the greatest resistance to forward motion comes from air resistance. The reduced friction means that maglev trains can move at very high speeds.

Figure 15.18 *The Maglev train in Shanghai started operating in 2002 and has reached speeds of 430 km/h.*

Two surfaces

Rough surfaces – greater frictional force

Smooth surfaces – less frictional force

Figure 15.19 *Friction acting between two surfaces.*

Friction in moving parts can be reduced by:

- using a lubricant such as oil or wax to create a layer between the two surfaces
- polishing surfaces that are in contact to make them as smooth as possible
- reducing the contact area between the moving parts.

GO! Activity 15.8

😊😊 **1. Experiment** In this activity you will investigate the friction acting on objects.

Read the plan carefully then follow the instructions, making sure you follow all of the safety rules as you would normally be expected to do when carrying out practical work.

You should work in a small group but record your own observations and complete the answers yourself.

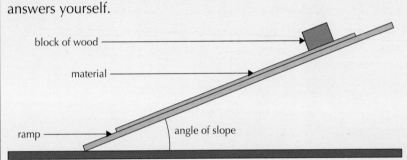

You will need: a ramp, a block of wood, a protractor and different materials (for example different grades of sand paper, felt, rubber, cling film, Vaseline jelly or vegetable oil).

Method

(a) Place a block on the ramp, with the ramp lying horizontal to the bench.

(b) Carefully lift the ramp at one end until the block begins to move.

(c) Hold the ramp in the position at the point where the block starts to move. Measure the angle of the slope with a protractor.

(d) Repeat this experiment at least three times and find the average angle of the slope at which the block begins to move.

(e) Change the surface, either by changing the material of the block or by changing the material covering the ramp, and repeat the experiment.

(f) Draw a suitable table to record the results.

(g) Which surface allows the block to slide the easiest? What conclusion can you draw from this experiment?

(h) If possible, try waxing, oiling or wetting the surface before lifting the ramp. Does this have an effect on the amount of friction acting between the surfaces?

? Did you know ...?

In 2014, the Great Britain curling teams won bronze and silver medals at the Winter Olympic Games. Curlers have to be able to move across the ice to sweep ahead of the curling stones. To make it easier to move, curlers wear shoes with different soles. A sliding sole on one foot is made of Teflon, which is very slippy. The other shoe has a rubber sole. This has more friction to help grip the ice, so the curlers can push with one foot and slide on the other foot.

Figure 15.20 *Curling stones on the ice.*

GO! Activity 15.9

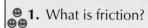

1. What is friction?

2. What are the units of friction?

3. What can be said about an object where the forces acting on it are balanced?

4. Think about a bike and all the places where you need friction. Can you think of other sporting activities that need friction? Discuss your thoughts with your group and write a short summary.

5. Is friction being increased or decreased in the following examples:

   (a) putting oil in a car engine

   (b) wearing tight-fitting swimwear

   (c) using an off-road bike with wider tyres.

Air resistance and drag

As a car moves along a road, it experiences a force opposing its forward motion from the air it is passing through. As the car moves, it has to push the air out of the way. The faster the vehicle is moving, the greater the force of friction. This is a type of friction called **air resistance**. **Drag** (water resistance) works in a similar way.

Gases and liquids contain particles. When an object travels through gas or air it collides with the particles. These collisions make it more difficult for the object to move through the fluid. A fast-moving object or a larger object collides with more particles and so the friction is greater.

When a car reaches its top speed, the air resistance is so large that it prevents the car going any faster. The pushing force from the engine is equal to the force of air resistance. The two forces are balanced.

Word bank

• **Air resistance**

Air resistance is the name given to the force of friction when objects travel through the air.

• **Drag**

Drag is the name given to the force of friction when objects travel through liquids.

Word bank

- **Streamlining**
Shapes are streamlined if they have a narrow, smooth shape that moves through a fluid easily.

Engineers use wind tunnels to investigate how the air flows over different shaped objects. Trails of smoke show how the air is flowing. The flow needs to be as smooth as possible. Where the flow has to change direction, sudden turbulence is created and this increases the drag. Reducing the area or **streamlining** the shape means the object meets less air and so will have lower frictional forces. (See Figure 15.21)

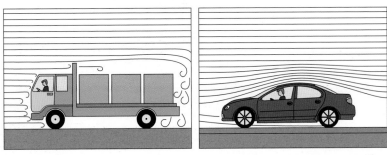

Figure 15.21 *Wind tunnels can be used in the design of cars to ensure they are a streamlined shape.*

GO! Activity 15.10

1. **Experiment** In this activity you will investigate the effect of an object's shape on its motion. You will drop objects of different shapes into a liquid and record the time the object takes to fall through the liquid.

Read the plan carefully then follow the instructions, making sure you follow all of the safety rules as you would normally be expected to do when carrying out practical work.

You should work in a small group but record your own observations and complete the answers yourself.

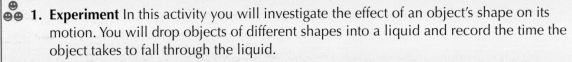

| flat disc | cylinder | cube | sphere | torpedo |

Possible shapes to investigate.

You will need: plasticine, a large measuring cylinder, a funnel, stopwatch and a large bowl.

Method

  (a) Using plasticine, make models of different shapes to test.
  (b) Fill a tall measuring cylinder with water. Use the funnel to help.
  (c) Hold the shape just above the surface of the water in the cylinder.
  (d) Drop the shape into the water and use a stopwatch to time how long it takes to reach the bottom.

(e) Repeat the experiment with different shapes.

(f) Record your results in a table.

(g) Of the shapes you tested, which one experienced the lowest frictional force?

(h) To streamline a car, designers try to give it a smooth shape. How could you change your experiment to test the design of cars?

Parachutes

Air resistance is used to slow parachutes as they fall to Earth and planes as they land at high speeds. Parachutes increase the surface area of a body as it falls through the air. This increases the amount of air that needs to be displaced.

? Did you know ...?

In July 2016, professional skydiver Luke Aikins became the first person to jump 25 000 feet from a plane without a parachute. He reached a speed of 120 mph when the forces acting on him became balanced. His downward force due to gravity was equal in size but opposite in direction to the friction force due to air resistance. This speed is called **terminal velocity**. He landed in a very large safety net. Most skydivers use a parachute to increase the surface area that is in contact with the air particles. This slows them down before landing.

GO! Activity 15.11

1. **Experiment** In this activity you will investigate the use of parachutes to slow the motion of falling objects.

Read the plan carefully then follow the instructions, making sure you follow all of the safety rules as you would normally be expected to do when carrying out practical work.

You should work in a small group but record your own observations and complete the answers yourself.

Making a plastic bag parachute.

You will need: a carrier bag, string or wool, scissors, a hole-punch, a small mass and a stopwatch.

(continued)

N3	L3	N4	L4

Method

(a) Cut a carrier bag into a large, flat square then trim the corners to make an octagon.

(b) Punch a small hole in the middle of the canopy. This helps the parachute to fall straighter.

(c) Use a hole-punch to punch small holes near each point of the octagon.

(d) Tie a piece of string or wool to each hole, then pull the other ends of those eight pieces of wool together and tie a knot.

(e) Drop your parachute with a small mass attached to the bottom of it and time how long it takes to fall to the ground. You could drop this down a stair-well or from a balcony. Should you repeat your test?

(f) Using the relationship between speed, distance travelled and time taken, calculate the speed of your falling mass: $v = \dfrac{d}{t}$

(g) Make a smaller parachute and repeat the experiment. What happens? Why do you think this is?

(h) What effect does the size of a parachute have on the speed of the falling object?

(i) If you have access to the right equipment, film the motion of your parachute and analyse it using computer software.

> ### Make the link
>
> Find out how to use the relationship between speed, distance and time in Chapter 14.

GO! Activity 15.12

😊😊 **1. Experiment** In this activity you will investigate the use of air-braking systems on moving vehicles.

Read the plan carefully then follow the instructions, making sure you follow all of the safety rules as you would normally be expected to do when carrying out practical work.

You should work in a small group but record your own observations and complete the answers yourself.

You will need: a ramp, books or a lab jack, a dynamics trolley or car, graph paper and card, straw, sticky tape and a tape measure or ruler.

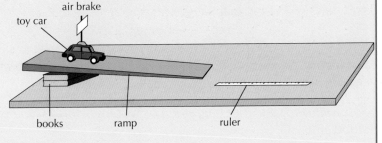

Method

(a) Set up the apparatus as shown in the diagram.

(b) Place the car on the top of the ramp and let it run as far as it goes until it stops.

(c) Measure and record the total distance from the bottom of the ramp to the front of the car with a tape measure or ruler. Repeat the experiment and calculate the average distance.

N3 L3 N4 L4

(d) Using the graph paper and card, create an 80 mm-tall air brake and attach it to the car.

(e) Repeat the experiment as before.

(f) Change the height of the air brake and continue collecting data.

(g) What kind of graph should you draw for these results? Draw the graph using your results.

(h) What conclusion can you draw from your results?

1. How could you make your experiment more accurate and reliable?

2. Search online to find information about air brakes on aeroplanes. How do the results of your experiment relate to the limited space for aircraft to land on aircraft carriers?

> **🔍 Hint**
>
> To calculate an average add all the test results together and then divide by the number of tests you performed.

Unbalanced forces

| National 3 |
| National 4 |
| Curriculum level 4 |
| Forces, electricity and waves: Forces SCN 4-07b |

Learning intentions

In this section you will:

- explain that objects can travel with constant acceleration when unbalanced forces are acting on them

- explain how changing the force or mass acting on an object changes its acceleration

- use the relationship between force, mass and acceleration.

The motion of an object is determined by the size and direction of the forces acting on it. In diagrams, forces are usually represented by an arrow, with the arrow showing the direction of the force.

If the forces acting on an object are not equal in size and direction, then the forces are **unbalanced**. An unbalanced force will cause a stationary object to start moving. An unbalanced force will also cause moving objects to change their speed (have constant acceleration) or change their direction. The acceleration of an object is proportional to the unbalanced force. If the unbalanced force increases, the acceleration will increase.

> **Word bank**
>
> - **Unbalanced force**
>
> Unbalanced forces occur when the forces acting on an object are not equal in size. Unbalanced forces cause objects to move or to accelerate.

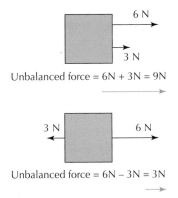

Unbalanced force = 6N + 3N = 9N

Unbalanced force = 6N – 3N = 3N

Figure 15.22 *Unbalanced forces.*

If two or more forces are acting on an object, the total unbalanced force is calculated by taking one away from the other. This is the **resultant force**.

Newton explained the effect of unbalanced forces in his Second Law.

Newton's Second Law

The acceleration of an object varies with the magnitude of the unbalanced force acting on the object and the mass of the object.

Example 15.1

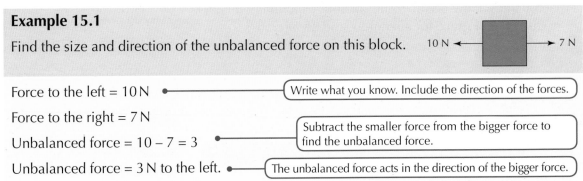

Find the size and direction of the unbalanced force on this block.

Force to the left = 10 N • ———— Write what you know. Include the direction of the forces.

Force to the right = 7 N

Unbalanced force = 10 – 7 = 3 • —— Subtract the smaller force from the bigger force to find the unbalanced force.

Unbalanced force = 3 N to the left. • —— The unbalanced force acts in the direction of the bigger force.

GO! Activity 15.13

☺ Which of the following boxes will accelerate? Work out the overall unbalanced force? Remember the arrows show the direction of the force.

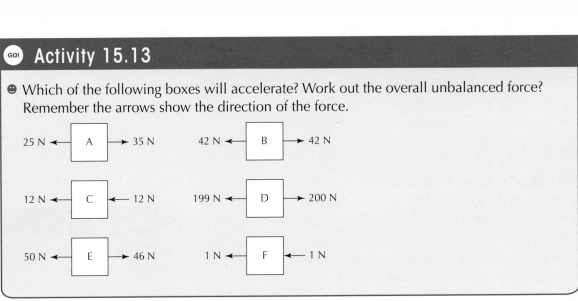

N3   L3   N4   L4

Activity 15.14

Watch a demonstration of the experiment shown in the figure below.

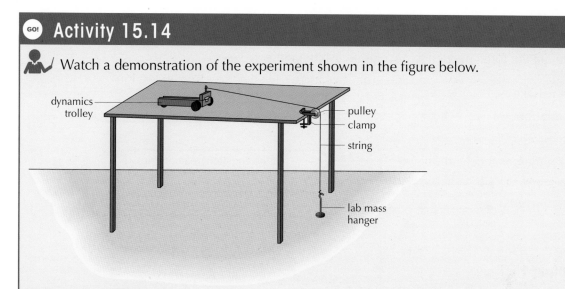

dynamics trolley
pulley
clamp
string
lab mass hanger

Investigating the effect of force on motion.

The trolley is initially at rest, but will start to move as the mass falls downwards. When the mass reaches the floor the string will go slack, but the trolley will keep moving.

1. What forces act on the trolley:
   (a) before it is moving?
   (b) while the mass is dropping?
   (c) once the mass has reached the floor?
2. Is the trolley accelerating (speeding up) or decelerating (slowing down)?
3. What would happen if the trolley were heavier?
4. What would happen if a larger force were pulling the trolley?

Calculations of force and acceleration

Newton's Second Law can be shown as the relationship shown here.

Force = mass × acceleration

This can be written as

$$F = ma$$

You will only have to use this relationship when the forces act in one direction.

Table 15.1 *Symbols and units used in the unbalanced forces relationship.*

Name	Symbol	Unit	Unit symbol
Force	F	newtons	N
Mass	m	kilograms	kg
Acceleration	a	metres per second squared	m/s^2

Example 15.2

Calculate the force needed to accelerate a 10 kg mass at 3 m/s².

mass = 10 kg

acceleration = 3 m/s² ———— Write out what you know from the question.

force = ?

force = mass × acceleration ———— Write out the equation (relationship).

force = 10 × 3 ———— Substitute in what you know.

force = 30 N ———— Don't forget to write the units in your answer.

Example 15.3

Calculate the acceleration of a 1200 kg mass when acted on by a force of 400 N.

m = 1200 kg

F = 400 N ———— Write out what you know from the question.

a = ?

$F = ma$ ———— Write out the equation (relationship).

400 = 1200 × a ———— Substitute in what you know and solve to find distance.

$a = \dfrac{400}{1200}$

a = 0.3 m/s² ———— Don't forget to write the units in your answer.

GO! Activity 15.15

☺ 1. Using the relationship for unbalanced force, complete the missing values **(a)–(f)** in the table.

Force (N)	Mass (kg)	Acceleration (m/s²)
(a)	5	10
(b)	10	0.4
(c)	12	15
5000	**(d)**	25
10 000	**(e)**	2.5
50	1225	**(f)**

2. Calculate the force needed to accelerate a 550 kg mass at 0.25 m/s².

3. Calculate the force needed to accelerate a 0.74 kg mass at 1.0 m/s².

4. Calculate the acceleration of a 10 kg mass object that is acted upon by a force of 40 N.

5. Calculate the acceleration of a 0.05 kg mass object that is acted upon by a force of 2 N.

6. Calculate the mass of an object that is accelerated at 5 m/s² by a force of 40 N.

7. Calculate the mass of an object that is accelerated at 20 m/s² by a force of 1040 N.

8. Give two reasons why a lorry will have a smaller acceleration than a small car.

The force of gravity

Curriculum level 3
Forces, electricity and waves: Forces SCN 3-08a
National 4

Learning intentions

In this section you will:

- explain the difference between mass and weight
- give examples of the effects of gravity on objects in different places in the Solar System
- use the relationship between weight, mass and gravitational field strength.

All objects on Earth experience the force of **gravity**, pulling them down towards the centre of the Earth. Gravity is always an attractive force.

Gravity acts between all objects, pulling them together. The strength of the attractive force is determined by the distance between the objects and the mass of the objects. The Earth experiences a force due to the gravity because of the Sun. Tides in the seas and oceans around the world are the result of the gravitational pull from the Moon exerted on the water in the oceans.

📖 Word bank

- **Gravity**

Gravity is the force pulling objects towards the Earth's centre.

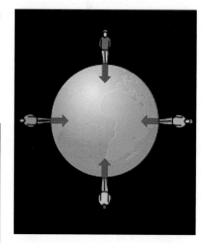

Figure 15.23 *The force of gravity acts all over the Earth towards the centre.*

❓ Did you know ...?

Galileo Galilei was an Italian scientist who lived in the sixteenth century. He is said to have dropped two spheres of different masses from the top of the Leaning Tower of Pisa to demonstrate that they would hit the ground at the same time. This contradicted the earlier thought that heavier objects would fall faster than lighter ones. The acceleration for each sphere is the same because of the gravitational pull of the Earth. Search online to find a video of the hammer and feather drop experiment that Apollo 15 astronaut David Scott performed on the Moon.

Figure 15.24 *The Leaning Tower of Pisa.*

Mass is a measure of the amount of material in an object. It is determined by the number and type of particles that make up the object. Mass does not depend on the force of gravity. It does not change if you move the object. If a can of baked beans has a mass of 0.4 kg on Earth it will have a mass of 0.4 kg on the Moon. Mass is measured in kilograms using a balance or a set of scales.

The **weight** of an object is the force acting on the object due to gravity. Weight is measured in newtons with a newton meter.

The term mass and weight are often mixed up. If there were no force due to gravity, objects would have a mass but be weightless.

📖 Word bank

- **Mass**

The mass of an object is a measure of the amount of matter in the object.

- **Weight**

The weight of an object is a measure of the force exerted on the object due to gravity.

Figure 15.25 *Measuring mass.*

GO! Activity 15.16

😊😊 **1.** Write a definition for mass.

2. Write a definition for weight.

3. Imagine a car crash on the Moon and the same car crash on Earth. There would probably be no difference in the damage between the two crashes. Explain why this is the case. Discuss your ideas with your group.

GO! Activity 15.17

😊😊 **1. Experiment** In this activity you will investigate the relationship between mass and weight.

Read the plan carefully then follow the instructions, making sure you follow all of the safety rules as you would normally be expected to do when carrying out practical work.

You should work in a small group but record your own observations and complete the answers yourself.

You will need: a clamp and stand, a force meter, a bag or polypocket, a pair of scales and a range of objects found in the lab or classroom such as a pencil case or a stapler.

Method

(a) Set up the apparatus as shown in the diagram.

(b) Measure and record the mass of an object using a set top balance or pair of scales. (Remember the units of mass.)

(c) Put this object into the bag hanging on the force meter and measure and record the weight. (Remember the units of weight.)

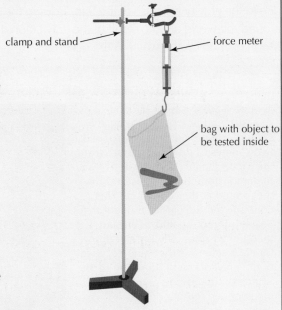

Experiment to find the relationship between mass and weight.

(d) Draw a suitable table to record your results.

(e) Repeat the experiment with different objects.

(f) Draw a graph of your results. (Hint: plot the mass on the x-axis and the weight on the y-axis.)

(g) Is there a relationship between the mass of an object and the weight? Is the graph a straight line?

(h) Graphs can be plotted using computers. The software will also find out information from the graph, such as the gradient. Try to find the gradient of your graph using a computer or other methods.

(i) Using the graph and your conclusion, suggest the weight of an object that has a mass of 200 g.

The **gravitational field strength** is the force of gravity acting on every kilogram of matter. It has units of newtons per kilogram (N/kg). The symbol for gravitational field strength is g.

The gravitational field strength on Earth is 9.8 N/kg. This means that 1 kg weighs 9.8 N. The strength of the force due to gravity is different on different planets. This is because other planets have different sizes and compositions. On large planets such as Jupiter, objects have a greater weight because there is a larger force due to gravity.

> ### 📖 Word bank
>
> • **Gravitational field strength**
> Gravitational field strength is the force of gravity acting on a kilogram of matter. Gravitational field strength depends on the size and mass of objects, and different planets have different field strengths.

Table 15.2 *The gravitational field strength on different planets.*

Planet	Mass of planet (10^{24} kg)	Gravitational field strength (N/kg)
Mercury	0.330	3.7
Venus	4.87	8.9
Earth	5.97	9.8
Mars	0.642	3.7
Jupiter	1898.0	23.0
Saturn	568.0	9.0
Uranus	86.8	8.7
Neptune	102.0	11.0

The force between bodies due to gravity is also affected by the distance between the bodies. If an object moves further away from a planet, it experiences a weaker gravitational force. Weightlessness occurs in outer space when an object is so far from any planet or a star that there is no noticeable force due to gravity.

Calculating weight

The **weight** of an object depends on the **mass** of the object and size of the **gravitational field strength**. We can use the following relationship to calculate the weight of an object:

> **Weight = mass × gravitational field strength**
>
> This can be written as
>
> $$W = mg$$

We can describe this relationship as 'weight is proportional to mass'.

Table 15.3 *Symbols and units used in the weight relationship.*

Name	Symbol	Unit	Unit symbol
Weight	W	newtons	N
Mass	m	kilograms	kg
Gravitational field strength	g	newtons per kilogram	N/kg

Example 15.4

Calculate the weight on Earth of a 30 kg chair (gravitational field strength g on Earth is 9.8 N/kg).

mass = 30 kg

gravitational field strength = 9.8 N/kg •————— [Write out what you know from the question.]

weight = ?

weight = mass × gravitational field strength •——— [Write out the equation.]

weight = 30 × 9.8 = 294 N •——————— [Substitute in what you know.]

└————————— [Don't forget to write the units in your answer.]

Example 15.5

Calculate the gravitational field strength on the Moon if a 25 kg mass has a weight of 40 N.

m = 25 kg

w = 40 N •————— [Write out what you know from the question.]

g = ?

$W = m \times g$ •————— [Write out the equation.]

$40 = 25 \times g$

$\dfrac{40}{25} = g$ •————— [Substitute in what you know.]

g = 1.6 N/kg •————— [Don't forget to write the units in your answer.]

? Did you know ...?

The value for gravitational field strength can also be used to describe the acceleration of an object falling towards the Earth.

GO! Activity 15.18

☻ **1.** Using the relationship for weight $W = mg$, complete the table for missing values **(a)–(i)**. Show your method for all the calculations.

Location	Weight (N)	Mass (kg)	Gravitational field strength (N/kg)
Mercury	(a)	75	3.7
Venus	(b)	5	8.9
Earth	98.0	10	(c)
Mars	(d)	3	3.7
Jupiter	2300.0	100	(e)
Saturn	603.0	(f)	9.0
Uranus	43.5	(g)	8.7
Neptune	(h)	2	11
Moon	(i)	150	1.6

Use the values of g given or calculated in the table above for the following questions.

2. State the mass of a 0.4 kg tin of beans on the Moon.

3. Calculate the weight of a 1 kg bag of sugar on the Moon.

4. Calculate the weight of a 0.5 kg object on Jupiter.

5. A 50 kg mass experiences a force due to gravity of 13500 N when close to the Sun. Calculate the gravitational field strength.

6. The dwarf planet Pluto has a mass of 0.177 times the mass of the Moon. Would you expect it to have a higher or lower gravitational field strength than the Moon?

GO! Activity 15.19

 1. **Experiment** Plan and carry out an experiment to investigate the relationship between the depth of craters and the force of gravity acting on a mass. You could drop marbles of different mass into trays of sand or flour. Think carefully about the other apparatus you will need and the measurements you will have to take. Include useful diagrams in your plan.

Make sure you follow all of the safety rules as you would normally do when carrying out practical work. You should work in a small group but record your own observations. Write a short report to share with your class. Don't forget to evaluate your experiment.

Force and energy

| National 3 |
| National 4 |

Learning intentions

In this section you will:

- describe the link between friction and heat energy
- learn that spacecraft need thermal protection systems.

When two surfaces rub together they get hot. If a block of wood is rubbed with sandpaper, the block of wood and the sandpaper will both get hot.

Moving objects have kinetic energy. When two objects move against each other, some of the kinetic energy of the movement is converted to heat energy because of friction. This heat energy can cause problems. For example, cylinders in a car engine heat up as they move inside the pistons.

To reduce this heating and avoid possible damage to the engine, special oils are used to lubricate the surfaces. The oil reduces the friction, and helps remove excess heat energy.

? Did you know ...?

Engine oils have to be highly engineered in order to work over a wide range of operating temperatures. They need to be able to work when the engine is starting from cold on a winter day and when the engine is running at a temperature of over 100°C.

Figure 15.26 *Shooting star.*

Figure 15.27 *A space shuttle docked with the International Space Station.*

Figure 15.29 *Space Shuttle Atlantis.*

Figure 15.30 *The black silica tiles on the bottom of a space shuttle.*

Space shuttle re-entry

When spacecraft re-enter the Earth's atmosphere there is a large amount of friction, as the spacecraft moves through the particles in the atmosphere. As the craft moves into the atmosphere, the particles oppose the motion. As the atmosphere gets thicker the force of friction increases. Some of the kinetic energy of movement is converted into heat energy.

The Soyuz descent modules which bring astronauts back from the International Space Station (ISS) reach speeds of hundreds of metres per second Temperatures on the outside of the module reach thousands of degrees Celsius. Without protection from the heat, the spacecraft would be destroyed.

The same effect can be seen with a shooting star. A shooting star is a meteor burning up as it travels through the Earth's atmosphere.

Space satellites and modules are protected from excess heat during space exploration. Descent modules returning from the ISS are protected by a heat shield made of composite materials. Most of the heat shield is vaporised on re-entry, while the module inside remains intact, protecting the astronauts inside. Parachutes slow the capsule down before landing.

? Did you know ...?

In the late 1960s and early 1970s, the American Apollo missions took 12 men to walk on the surface of the Moon. The Apollo 13 mission to the Moon was interrupted when there was an explosion on board the spacecraft. The crew made vital repairs and returned to Earth safely. The film *Apollo 13* is based on these events.

Figure 15.28 *Apollo 13 Command Module.*

In the 1980s, NASA started a new programme of space missions using space shuttles. Unlike previous missions, the space shuttles were designed to be reusable, and were capable of landing like a plane on their return from space.

Between 1982 and 2011, 135 space shuttle missions were flown carrying out tasks such as taking astronauts to the Spacelab space station, retrieving satellites and carrying out repairs to the Hubble Space Telescope.

The underside of the shuttles were covered in special tiles made from a black silica compound. The tiles transfer the heat back into the atmosphere very quickly.

Activity 15.20

1. **Experiment** In this activity you will investigate the effect of colour on the rate of heat transfer. Why are the special tiles on the bottom of a space shuttle painted black and not white?

Read the plan carefully then follow the instructions, making sure you follow all of the safety rules as you would normally be expected to do when carrying out practical work.

You should work in a small group but record your own observations and complete the answers yourself.

You will need: stopwatch, three thermometers, three different coloured containers with lids (lids must have holes for the thermometers, containers must be the same size, shape and material), a kettle.

Method

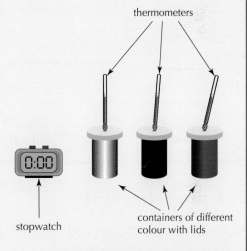

thermometers

stopwatch

containers of different colour with lids

(a) Set up the apparatus as shown in the diagram.

(b) Pour hot water into each container. Use a measuring cylinder to measure the same volume of water each time. Take care when using hot water!

(c) Place the lid on the container and record the initial temperatures.

(d) After 1 minute, measure the temperature again. Which container has lost the most heat energy?

(e) After 5 minutes, measure the temperature again.

(f) Which colour of container lost the most heat energy during this experiment?

(g) Why do you think the containers needed lids?

(h) What colour are the tiles covering the space shuttles? Do the results of this experiment link to this colour?

◌ Make the link

The way different colours transfer heat energy is discussed in more detail in Chapter 8.

? Did you know ...?

Space Shuttle Columbia disintegrated as it re-entered the Earth's atmosphere in February 2003. All seven crew members were killed. A piece of insulation designed to protect the shuttle from extreme temperatures broke off and struck the left wing. The wing was destroyed, causing the space shuttle to become unstable and break apart.

🔄 Keep up to date!

Many people, scientists and governments think that space exploration is worth all the risks and expense involved. Our understanding of the Universe, the information learned about the Earth and the scientific developments gained have benefited everyday life. Search online to find out what technologies used in everyday life have come from space exploration.

GO! Activity 15.21

☺ **1.** Copy and complete the following paragraphs about friction and heat energy, selecting the correct word from the options provided.

Moving objects have _____(**kinetic/heat**) energy. When two objects move against each other, some of the energy is converted to _____(**heat/kinetic**) energy because of _____(**gravity/friction**). This energy can cause problems. _____(**Water/Oil**) can be used to lubricate the moving parts to help _____(**reduce/increase**) the force of friction.

Space shuttles and modules have to be protected from excess heat during _____ (**take-off/re-entry**) through the Earth's atmosphere. The silica tiles on the bottom of the space shuttle are _____ (**black/white**) as this colour is a good _____ (**conductor/insulator**) of heat energy.

Learning checklist

After reading this chapter and completing the activities, I can:

`N3` `L3` `N4` `L4`

- describe how forces are measured with a force meter. **Activity 15.2, Activity 15.3** ⚪ ⚪ ⚪

- state that forces change the shape, speed or the direction of moving objects. **Activity 15.4, Activity 15.5 Q1** ⚪ ⚪ ⚪

- give examples of different effects and the application of forces. **Activity 15.4** ⚪ ⚪ ⚪

- explain how measurements of forces are used to keep us safe. **Activity 15.6, Activity 15.7** ⚪ ⚪ ⚪

- explain how objects can travel at constant speed when balanced forces are acting. **Activity 15.8 Q3** ⚪ ⚪ ⚪

- state that friction is a force that acts to resist movement. **Activity 15.8 Q1** ⚪ ⚪ ⚪

- describe how to measure the force of friction. **Activity 15.9, Activity 15.10** ⚪ ⚪ ⚪

N3 L3 N4 L4

- give example of places where friction is wanted and not wanted. **Activity 15.8 Q4, Q5, Activity 15.11, Activity 15.12** ○ ○ ○

- explain that objects can travel with constant acceleration when unbalanced forces are acting. **Activity 15.13** ○ ○ ○

- explain how changing the force or mass acting on an object changes the object's acceleration. **Activity 15.14** ○ ○ ○

- use the relationship between force, mass and acceleration. **Activity 15.15** ○ ○ ○

- explain the difference between mass and weight. **Activity 15.16 Q1, Q2, Activity 15.17** ○ ○ ○

- give examples of the effects of gravity on objects in different places in the Solar System. **Activity 15.16 Q3** ○ ○ ○

- use the relationship between weight, mass and gravitational field strength. **Activity 15.18** ○ ○ ○

- describe the link between friction and heat energy. **Activity 15.21** ○ ○ ○

- describe the thermal protection systems used on spacecraft. **Activity 15.21** ○ ○ ○

16 Satellites

This chapter includes coverage of:

N4 Satellites

You should already know:

- that any object which orbits a planet or a star is called a satellite
- that the Moon is a natural satellite orbiting the Earth
- that the Earth is a natural satellite orbiting the Sun.

National 4

Satellite orbits

Learning intentions

In this section you will:

- identify examples of natural satellites
- learn that the function of an artificial satellite depends on the height of its orbit above the Earth
- give examples of the functions of satellites in different orbits around the Earth
- learn that the period of a satellite depends on the altitude of the satellite above the Earth
- learn that when viewed from the Earth, a geostationary satellite stays at a fixed position in the sky
- learn that a geostationary satellite has a period of 24 hours.

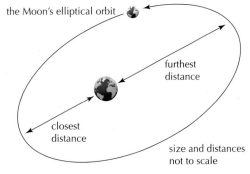

the Moon's elliptical orbit

furthest distance

closest distance

size and distances not to scale

Figure 16.1 *The distance between the Moon and the Earth is constantly changing, due to its elliptical orbit. Super moons occur when the Moon is closest to the Earth.*

A satellite is an object in **orbit** around a planet or star. The Moon is a **natural satellite**, orbiting around the Earth. The Earth is itself a satellite, orbiting around the Sun. Both the Moon and the Earth follow elliptical orbits. This means that the orbits are not quite circular, but are oval-shaped. The distance from the Earth to the Moon varies from 363 396 km to 405 504 km.

An artificial satellite is an object put into orbit around the Earth by humans. The first artificial satellite, called Sputnik 1, was launched by the Soviet Union in October 1957.

N3 | L3 | **N4** | L4

Since then nearly 7000 satellites have been launched. There are approximately 1300 functioning satellites orbiting the Earth.

The function of a satellite depends on its height above the Earth's surface.

Most satellites orbit at three height ranges above the Earth's surface:

- low Earth orbit (LEO)
- medium Earth orbit (MEO)
- geostationary orbit (GEO).

Table 16.1 shows the numbers of satellites in these orbits and their height and the **period** of the orbit. The table shows that the period of the orbit is related to the height of the orbit.

Table 16.1 *Height and period of satellite orbits*

Orbit	Approximate number of satellites	Height above Earth's surface (km)	Period of satellite (hours)
Low Earth orbit	500	180–2000	1.5 (90 minutes)
Medium Earth orbit	50	2000–36 000	12
Geostationary orbit	500	36 000	24

Low Earth orbiting satellites

Most satellites orbiting the Earth are in low Earth orbits. They orbit with a period of approximately 90 minutes. Most scientific and observation satellites used for monitoring the environment are in this orbit. They are used for detecting changes in the Earth's vegetation, monitoring the gas chemical concentrations in the atmosphere, or mapping terrain.

Low Earth orbiting satellites can be used for imaging the movement of glaciers or mapping buildings destroyed in natural disasters, such as earthquakes. Spy satellites (also called reconnaissance satellites) orbit at this height. The relatively low height of orbit means they can see the surface much more clearly.

Polar orbiting satellites pass above the north and south poles on each revolution of the Earth, orbiting in 90 minutes. They are used for Earth monitoring and weather observation. As the Earth rotates, they pass over the entire surface in the course of a full day.

📖 Word bank

- **Orbit**

An orbit is the curved path of an object as it revolves around a planet or star. Objects stay in orbit due to gravity.

📖 Word bank

- **Period**

The time taken for a satellite to complete a single orbit is its period.

🔍 Hint

The higher a satellite orbits the Earth, the greater the period of orbit.

Figure 16.2 *Satellite image of the Fukushima nuclear power plant, in Japan, after the 2011 earthquake and tsunami.*

Figure 16.3 *Polar orbiting satellites scan the entire Earth as it rotates.*

✦ Physics in action: Scientific endeavours in low Earth orbits

The Hubble Space Telescope was launched into a low Earth orbit in 1990. It can take images in the ultraviolet, visible and infrared regions of the EM spectrum. The telescope is over 13 metres long, and the mirror used to capture the light is 2.4 m in diameter.

The Hubble telescope has been responsible for some significant discoveries. It has helped to improve estimates for the age of the universe, and has taken some iconic images. The Hubble Ultra Deep Field image, in Figure 16.6, has revealed galaxies billions of light years away.

The most well-known image from the Hubble Space Telescope is the Eagle Nebula's Pillars of Creation (Figure 16.7).

Figure 16.7 *The Pillars of Creation.*

The nebula shows stars forming in a region of space so far away that light from it takes 7000 years to reach us.

The International Space Station (ISS) is a habitable artificial satellite in a low Earth orbit. It was launched in 1998 and has had astronauts living on board since 2000. It is used as a space research laboratory for experiments in all areas of science.

The British astronaut Tim Peake spent six months on the ISS from December 2015 to June 2016. He created many videos explaining what life in space is like. He also became the first man to run a marathon in space.

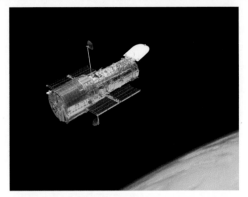

Figure 16.5 *The Hubble Space Telescope orbiting the Earth.*

Figure 16.6 *The Hubble Ultra Deep Field image. Each dot of light in this image is a galaxy. The Hubble Space Telescope observed a seemingly empty patch of sky for one million seconds, and detected all these galaxies.*

Figure 16.8 *Tim Peake preparing to install a sensor to measure the forces on the space station due to crew activities, docking and manoeuvring.*

N3 | L3 | **N4** | L4

Many artificial satellites can be seen from the surface of the Earth in the night sky. Most look like stars, reflecting the light from the Sun as they pass hundreds of kilometres overhead, so they can be hard to spot in the sky. Unlike stars, many satellites move across the sky at night quite quickly. There are many smartphone apps that help to spot and track satellites – why not use one to try to find some satellites on a clear night?

Figure 16.4 *A smartphone app tracking live satellites over the British Isles.*

Medium Earth orbiting satellites

Most satellites in medium Earth orbits are at an altitude of around 20000 km above the Earth's surface. They have a period of 12 hours. Satellites in this region are mostly used for navigation and communication. The Global Positioning System (GPS), originally developed by the USA, uses 32 satellites orbiting at this altitude to provide accurate positioning information. The orbits of each satellite are arranged so that at least six satellites are visible from any point on the Earth's surface at any moment.

The GPS programme was originally developed for American military use. It is now available for anyone to use. You can use GPS apps on your smartphone, and many cars have an inbuilt SatNav (`satellite navigation') system which uses GPS. Other countries have their own positioning satellites systems. Russia has a network of 24 satellites in their GLONASS system. The European Space Agency (ESA) is developing a network of 24 satellites for their Galileo system.

Figure 16.9 *GPS has improved navigation in unfamiliar places.*

Geostationary orbiting satellites

Satellites that stay at the same position above the Earth's surface are called geostationary satellites. When viewed from the Earth, they remain fixed in one location in the sky. They are, however, orbiting at the same rate as the Earth's rotation, and in the same direction. All geostationary satellites orbit at the same height (around 36000 km) and have a period of rotation of 24 hours.

Geostationary satellites are used for communications and for weather monitoring. The main advantage of a geostationary satellite is its fixed position in the sky. This means that

| N3 | L3 | **N4** | L4 |

transmitting and receiving antennas on Earth do not need to track the satellite, but can simply be fixed in place to point in one direction. If you look at a block of flats with TV satellite dishes, you will see they are all pointing in the same direction.

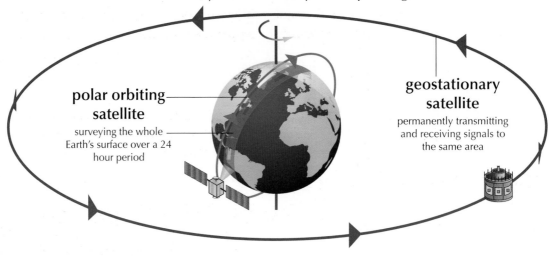

polar orbiting satellite
surveying the whole Earth's surface over a 24 hour period

geostationary satellite
permanently transmitting and receiving signals to the same area

Figure 16.10 *Comparison of polar orbiting and geostationary satellites.*

📖 Word bank

• **Space junk**
Space junk, or debris, comprises any large or small piece of junk orbiting the Earth. Space junk can be both natural and man-made.

🔄 Keep up to date!

The 2013 movie *Gravity* explores the issue of space junk orbiting the Earth. **Space junk** can be bits of old rockets such as spent booster rockets used to put satellites in orbit, non-functioning satellites, or a variety of smaller items used in space.

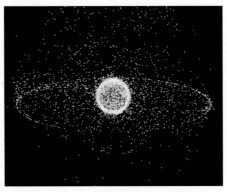

Figure 16.11 *An image taken by a NASA satellite in a high Earth orbit of space debris. The wide ring is debris in a geostationary orbit, and the inner ring close to the Earth is debris in a low Earth orbit.*

Search online to investigate the risks that orbiting space junk impose on space exploration. Think about what aspects of modern life could be impacted if space junk is allowed to accumulate as we continue to use satellites around the Earth.

GO! Activity 16.1

☺ 1. The Moon is a satellite, since it orbits the Earth.

    (a) State whether the Moon is an artificial or natural satellite.

    (b) State another example of the same type of satellite.

2. Artificial satellites orbit at different heights. Give an example of the type of satellite that might orbit at:

    (a) a low Earth orbit

    (b) a medium Earth orbit

    (c) a geostationary orbit.

3. The Hubble Space Telescope orbits at an altitude of 547 km and the International Space Station (ISS) orbits at an altitude of 400 km.

    (a) Both these satellites are in low Earth orbits. What is the range of heights that satellites can be in low Earth orbits?

    (b) Which satellite, the Hubble Space Telescope or the ISS, takes longer to orbit the Earth? Explain your answer.

4. Many TV networks use geostationary satellites for communication.

    (a) State three differences between geostationary satellites and polar orbiting satellites.

    (b) Explain why geostationary satellites are more suitable for transmitting satellite TV.

Communicating with satellites

National 4

Learning intentions

In this section you will:

- learn how to use the relationship between distance, speed and time to solve problems on satellite communication
- explain how curved reflectors are used in satellites to send and receive signals.

☀ Make the link

Radio waves and microwaves are low-energy waves in the electromagnetic spectrum. Find out more about the electromagnetic spectrum in Chapter 12.

Satellites communicate with ground-based transmitters and receivers using electromagnetic waves. Radio waves and microwaves are most commonly used in satellite communication.

Orbiting satellites require a power source to operate. They are equipped with rechargeable batteries which are charged using large solar panels. The panels convert light from the Sun into electrical energy.

Figure 16.12 *Solar cells on orbiting satellites convert light energy into electrical energy.*

Word bank

- **Line-of-sight communication**

Line-of-sight communication takes place when there is a straight, direct path between the transmitter and receiver.

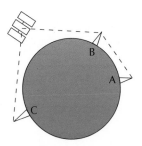

Figure 16.13 *Signals can be sent between A and B, but transmitter C cannot send signals to either A or B because of the curvature of the Earth.*

Satellite communication

For many types of communication which use signals sent through the air, the transmitter and receiver must be in the **line of sight**. When communication is required over a long distance, there is no direct line of sight because of the curvature of the Earth.

Satellites in orbit are used in long-distance communications. A satellite is positioned in orbit in direct line of sight of both transmitter and receiver. The satellite receives the signal from the transmitter and re-transmits it to the receiver.

The time taken for the signal to reach the receiver depends on the distance the signal has to travel and the speed of the signal:

$$d = vt$$

where:
d = **distance** (that the signal travels)
v = **speed** (of the signal)
t = **time** (for the signal to travel.

We can use the equation given above to work out how long it would take a signal to be sent from one transmitter on the ground, via satellite, to a receiver a long distance away. All radio or microwave signals used in satellite communication travel at the speed of light, $v = 300\,000\,000$ m/s. (Remember the speed of light can also be written as 3×10^8 ms^{-1}.)

Example 16.1

Television pictures from the 2016 Rio Olympics were sent from Rio to London using a satellite. The signals were sent using microwaves. The distance between Rio and the satellite is 36 200 km. The distance from the satellite to London is 36 800 km.

Calculate the time taken for the signal to travel from Rio to London. Assume the signal is re-transmitted instantaneously by the satellite.

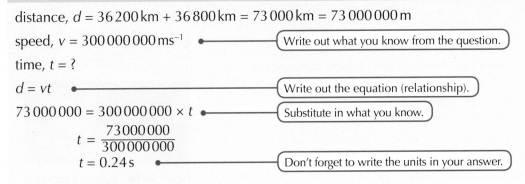

distance, $d = 36\,200$ km + 36 800 km = 73 000 km = 73 000 000 m

speed, $v = 300\,000\,000$ ms^{-1} ⟵ Write out what you know from the question.

time, $t = ?$

$d = vt$ ⟵ Write out the equation (relationship).

$73\,000\,000 = 300\,000\,000 \times t$ ⟵ Substitute in what you know.

$t = \dfrac{73\,000\,000}{300\,000\,000}$

$t = 0.24$ s ⟵ Don't forget to write the units in your answer.

Activity 16.2

☺ **1.** Calculate the time taken for a radio signal to travel:
   (a) the length of the United Kingdom, 1400 km
   (b) around the world once, a distance of 40 000 km
   (c) to the Moon and back, 769 000 km.

2. Calculate the distance, in km, to the International Space Station, if it takes 0.0013 s for a radio signal to travel between the Earth and the space station.

3. The Solar and Heliospheric Observatory is a satellite used by NASA and the ESA to monitor the Sun. It orbits at 1.5 million km away from the Earth. Calculate how long it takes radio signals to travel from the satellite back to Earth.

Curved reflectors

Many houses in Scotland have satellite dishes on the exterior wall or roof to receive satellite TV signals from an orbiting communication satellite.

These signals come from satellites in geostationary orbits, high above the Earth's surface. The signals are very weak because of the large distances the signals travel. The dishes are curved in a parabolic shape, which collects the signal and reflects it to a focus. An antenna at the focus picks up a much stronger signal than it would without the **curved reflector**.

Figure 16.14 *Satellite dishes for receiving TV signals.*

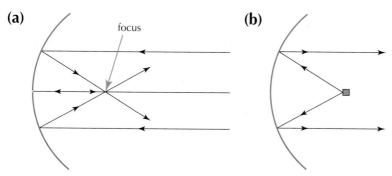

Figure 16.15 **(a)** *A curved receiving dish collects weak signals and reflects them to a focus, giving a much stronger signal.* **(b)** *A curved transmitting dish reflects rays to produce a parallel beam. Notice the difference between the direction of the signal between **(a)** the receiving dish and **(b)** the transmitting dish.*

Curved reflectors are used in radio telescopes, solar cookers and many other applications.

Curved reflecting dishes are also used for transmitting signals. A transmitter placed at the focus sends signals towards the dish, which are reflected to produce a parallel beam. This signal can be directed precisely and transmitted over a long distance.

Make the link

Curved reflectors are investigated in the 'Reflection of light' section of Chapter 11. They behave in the same way as concave mirrors.

Figure 16.16 *Curved transmitter dishes.*

GO! Activity 16.3

😊😊 1. **Experiment** In this activity, you will investigate how curved reflectors improve the signal received from a microwave transmission.

Read the plan carefully then follow the instructions, making sure to follow all of the safety rules as you would normally be expected to do when carrying out practical work.

You can work in small groups but you should record your own observations and complete the answers yourself.

You will need: a curved reflector dish, power supply and microwave transmitter, microwave receiver, a lamp jack or clamp stand (x2).

Method

(a) Set up the transmitter and receiver so that they are within 5 cm of each other. Position them on clamp stands (or lab jacks) as shown.

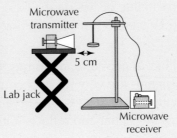

(b) Turn on the transmitter and receiver and ensure that you can receive a strong signal from the microwave transmitter.

(c) Move the receiver away from the transmitter until the signal strength is reduced to close to zero. This distance will range from 15 to 30 cm depending on the strength of the microwaves transmitted.

(d) Now position a curved reflecting dish behind the receiver. Position the dish carefully so that the microwave receiver is at the focus point of the curved reflector. What happens to the strength of the signal received?

🔍 Hint

Curved reflectors have two functions in communications:
- as **receiving** dishes, to collect weak radio signals
- as **transmitting** dishes, to send signals in a parallel beam.

Orbiting satellites will often have both transmitting and receiving curved reflectors.

(e) Draw a ray diagram to show how the curved reflector increases the strength of signal received by the microwave receiver.

🔍 Hint

Look back at Figure 16.15.

| N3 | L3 | **N4** | L4 |

GO! Activity 16.4

☺ **1.** A GPS satellite is directly above the surface of the Earth. It orbits at a height of 20 000 km. How long does it take a signal to travel from the satellite to the GPS receiver on Earth?

2. A satellite signal is sent from a sporting event to another country across the globe using an orbiting satellite. The satellite is 6000 km from each receiving station on Earth, as shown.

6000 km   6000 km

(a) Calculate the total distance the signal has to travel between each country.

(b) How long does it take for the signal to travel from one country to the other?

3. Curved reflectors are commonly used to receive signals from satellites in orbit. This figure shows a diagram with three rays reflecting off a curved reflector.

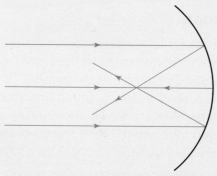

(a) State the name given to the position where the reflected rays intersect.

(b) Explain why a receiving antenna placed at this point would receive a stronger signal than if the dish were not there.

4. Reflectors have been placed on the Moon during several of the Apollo Moon landings, to help accurately determine the distance to the Moon. In one measurement, the light from a laser beam directed at a reflector on the Moon takes 2.5 s to travel to the Moon and back again. The laser beam travels at 300 000 000 m/s.

(a) Calculate the distance travelled by the laser beam.

(b) Using your answer to part **(a)**, determine the distance to the Moon.

Uses of satellites

Learning intentions

In this section you will:

- describe applications of satellites, including telecommunications, navigation, military observation, weather monitoring and environmental monitoring
- learn how satellites are developing our understanding of the global impact of mankind's actions.

Artificial satellites have a significant positive impact on our lives, although we rarely think about them. Satellites are placed into orbit for:

- telecommunications
- navigation
- military observation
- weather monitoring
- environmental monitoring.

Telecommunications

Before the use of satellites for television communication, TV signals could only travel short distances. In 1964, the satellite Syncom 3 was the first satellite to be used to transmit TV signals of the Olympics Games in Tokyo. Communication satellites in geostationary orbits are now commonly used to transmit TV signals around the world. News reports of world events can be communicated using satellites as they happen.

Satellites are used to communicate in remote areas where there are no fixed landline communications, such as in mountain ranges, deserts and oceans. In these locations, satellites supply voice and internet services.

Figure 16.17 *Satellite phones and messengers are used by hillwalkers in remote areas of Scotland, where there is no mobile phone signal coverage.*

N3 | L3 | **N4** | L4

Navigation

Global positioning systems (GPS) use a network of satellites to provide accurate location and navigation information. If a GPS receiver can determine the distance to at least three GPS satellites, then its location can be pinpointed on the Earth's surface.

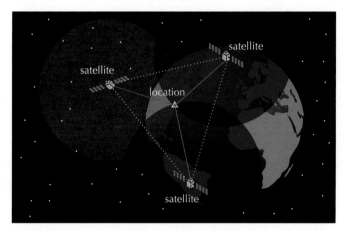

Figure 16.18 *GPS satellites can determine location accurately.*

Military observation

Satellites used for military or intelligence applications are known as **reconnaissance satellites**. These satellites can be used for:

- missile early-warning systems

- detecting nuclear explosions

- photo surveillance

- intercepting radio signals for intelligence uses

- detecting objects at night using radar.

Reconnaissance satellites help to protect countries through surveillance, and are also used to show evidence of military action and attacks in war zones.

Weather monitoring

Weather satellites provide information on cloud cover, storms, rain levels and temperature. They observe light from both the visible and infrared regions of the electromagnetic spectrum. The visible light provides information about cloud cover. Analysis of the infrared light is used to calculate the presence of water vapour in the atmosphere. Infrared analysis is also used to measure land and sea surface temperatures.

? Did you know ...?

The UK has its own range of reconnaissance satellites, called Skynet.

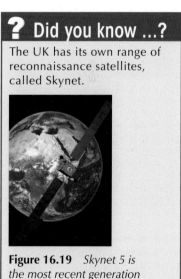

Figure 16.19 *Skynet 5 is the most recent generation of satellites used by the UK.*

Environmental monitoring

Satellites are also used to monitor changes in the Earth's environment. They can observe the Earth's vegetation, pollution levels in the atmosphere, ocean temperatures and ice field movements. Satellites can examine areas of deforestation and regions of drought. They can also detect eruptions from volcanoes and the resulting motion of ash clouds.

↻ Keep up to date!

The European Space Agency (ESA) is running an ambitious Earth-monitoring programme called Copernicus. The Copernicus programme consists of seven LEO satellite missions, with each mission using at least two satellites. The programme is collecting a huge amount of data from land, oceans and the atmosphere to provide information which will help in planning services and applications, and to help give a better understanding of the impact that human activities are having on the environment.

In August 2016, one of the Sentinel 1A Earth monitoring satellites captured images of ground movement after a devastating earthquake in Italy. A few days earlier the same satellite had been hit with space debris, causing damage to one of its solar panels.

Figure 16.21 *Scientists working on the Sentinel 2A, prior to launch in June 2015. It monitors farming and forestry practices, and helps to manage food security.*

Use the ESA website to find out about the latest developments with the Copernicus programme, and about the challenges the programme faces in using satellites in low Earth orbits.

☺ Activity 16.5

☻ 1. For each of the following types of satellite, describe one way in which it has benefitted society:

   (a) military satellite

   (b) navigation satellite

   (c) weather satellite.

2. Environmental satellites can monitor levels of greenhouse gases that contribute to climate change. State two other ways these satellites can provide information on mankind's impact on the planet.

Learning checklist

After reading this chapter and completing the activities, I can:

N3 L3 **N4** L4

- state that the Moon and the Earth are examples of natural satellites. **Activity 16.1 Q1** ○ ○ ○

- state that the function of an artificial satellite depends on the height of its orbit above the Earth. **Activity 16.1 Q2** ○ ○ ○

- give examples of the function of satellites at different orbits around the earth. **Activity 16.1 Q2** ○ ○ ○

- state that the period of a satellite depends on the altitude of the satellite above the Earth. **Activity 16.1 Q3** ○ ○ ○

- state that a geostationary satellite stays at a fixed position above the Earth. **Activity 16.1 Q4** ○ ○ ○

- state that a geostationary satellite has a period of 24 hours. **Activity 16.1 Q4** ○ ○ ○

- use the relationship between distance, speed and time to solve problems on satellite communication. **Activity 16.2, Activity 16.4 Q1, Q2, Q4** ○ ○ ○

- explain how curved reflectors are used in satellites to send and receive signals. **Activity 16.3, Activity 16.4 Q3** ○ ○ ○

- describe applications of satellites, including telecommunications, navigation, military observation, weather and environmental monitoring. **Activity 16.5 Q1** ○ ○ ○

- explain how satellites are developing our understanding of the global impact of mankind's actions. **Activity 16.5 Q2** ○ ○ ○

17 Planet Earth and the Solar System

You should already know:

- that the Sun is a star at the centre of our Solar System
- that our Solar System has eight planets orbiting around the Sun
- that objects which orbit around a planet or star are called satellites
- that the Moon is a natural satellite of the Earth, and the Earth is a natural satellite of the Sun
- how lunar and solar eclipses occur.

National 3

Our Planet Earth

Learning intentions

In this section you will:

- identify the different layers of the Earth
- investigate different types of rock
- learn what is meant by 'the rock cycle'
- learn that the shifting of tectonic plates can cause earthquakes
- explore how volcanoes are caused.

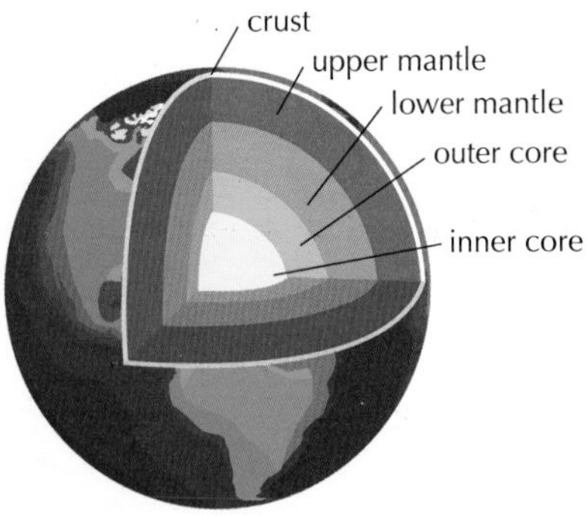

crust
upper mantle
lower mantle
outer core
inner core

Figure 17.1 *The structure of the Earth.*

Our planet is the only planet orbiting the Sun that contains life (that we know of). Humans have developed an understanding of many features of planet Earth.

The Earth is constructed of four distinct layers. These are the:

- crust
- mantle
- outer core
- inner core.

Table 17.1 *The layers of the Earth.*

Layer	Details	Thickness (km)	Temperature (°C)
Crust	The crust is the outermost layer of the Earth. It is rocky and thin compared to the other layers. It is thinnest beneath the oceans.	8–70	Around 22
Mantle	The mantle is the thickest layer. It is made of semi-molten rock called magma. The upper mantle is harder than the lower mantle, because it is cooler. The lower mantle is very hot. It would melt if it were not for the large pressure pushing down on it.	2900	Between 1400 and 3000
Outer core	The outer core is a liquid layer of molten iron and nickel. It flows around the inner core and causes the Earth's magnetic field.	2200	4000–6000
Inner core	The inner core is the centre of the Earth. It is a solid ball of iron and nickel. It is very hot, but remains solid because of the enormous pressures around it.	1300 (radius)	5000–6000

The rocks of the Earth's crust

The Earth's crust is made of three types of rocks:

- igneous
- sedimentary
- metamorphic.

Igneous rocks make up over 90% of the Earth's crust. Igneous rock is formed from **magma** from the mantle of the Earth. Basalt and granite are examples of igneous rock. Most igneous rocks are very hard. The oldest rocks in Scotland are igneous rocks which form most of the islands of the Outer Hebrides.

Scottish mountains such as Ben Nevis and the Cairngorms are made of granite.

Sedimentary rock is formed from matter that has been laid down on the Earth's surface, or in oceans and other bodies of water such as lakes, and has been compressed over time. The pressure causes the matter to turn to rock. These rocks cover around 80% of the Earth's land area, but make up only a small amount of the Earth's crust.

Common sedimentary rocks are limestone, sandstone, chalk and clay. Most sedimentary rocks are quite soft compared with igneous rocks.

Sandstone is often used as a building material, and many of the old buildings in Edinburgh and Glasgow were built using sandstone.

? Did you know ...?

The inner core of the Earth is about ¾ the size of the Moon, but is as hot as the surface temperature of the Sun!

Figure 17.2 *Fingal's Cave on the uninhabited island of Staffa, in the Inner Hebrides of Scotland, is formed entirely of basalt columns.*

📖 Word bank

- **Magma**

Magma is molten or semi-molten rock that is found beneath the Earth's crust.

Figure 17.3 *Many of the buildings in Aberdeen, such as the City Council building shown here, are made of granite.*

Metamorphic rock is formed from rock that has been changed by extreme heat and pressure into another type of rock. For example, marble is formed from limestone, and slate is formed from mudstone or shale. Marble and slate are both types of metamorphic rock.

Metamorphic rocks are very hard-wearing. Slate is often used in the building industry as a roofing material. Some of the oldest rocks in Earth are metamorphic, and are found in the Outer Hebrides of Scotland.

The west coast of Scotland had many slate quarries in the eighteenth and nineteenth centuries.

Figure 17.4 *Sedimentary rock on the Isle of Arran.*

Figure 17.5 *Slate is an ideal roofing material.*

Word bank

• **Erosion**

Erosion is the wearing away of soil, rock or other materials, often by water, and the transportation of that material to another location.

The rock cycle

The rocks of the Earth are continually changing in the rock cycle. Igneous and metamorphic rocks lying on the surface are exposed to the weather and worn away in a process called **erosion**. Rivers carry the erosion products such as sand, mud and pebbles to the sea, to form sedimentary rock. Movement of the Earth's crust can cause these sedimentary rocks to undergo great extremes of pressure and temperature, to form metamorphic rocks.

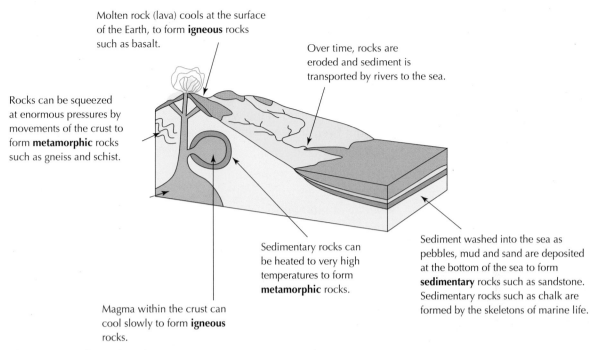

Molten rock (lava) cools at the surface of the Earth, to form **igneous** rocks such as basalt.

Over time, rocks are eroded and sediment is transported by rivers to the sea.

Rocks can be squeezed at enormous pressures by movements of the crust to form **metamorphic** rocks such as gneiss and schist.

Sediment washed into the sea as pebbles, mud and sand are deposited at the bottom of the sea to form **sedimentary** rocks such as sandstone. Sedimentary rocks such as chalk are formed by the skeletons of marine life.

Sedimentary rocks can be heated to very high temperatures to form **metamorphic** rocks.

Magma within the crust can cool slowly to form **igneous** rocks.

Figure 17.6   *The rock cycle.*

☀ Physics in action: Plate tectonics

The Earth's crust is made up of huge plates that move very slowly. The movement of these plates is called plate tectonics. The theory of plate tectonics was developed by Arthur Holmes, a geologist at Edinburgh University in the late 1920s.

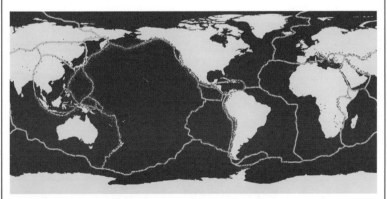

Figure 17.8   *The Earth's tectonic plate boundaries are shown in blue. Areas of high earthquake activity are shown in red.*

The movement of the plates is caused by convection currents in the Earth's mantle. The movement of the plates is very slow.

(*continued*)

❓ Did you know ...?

The concept of the rock cycle is credited to a Scottish doctor named James Hutton (1726–1797). He is often referred to as the father of modern geology.

Figure 17.7   *James Hutton.*

Mountain ranges are formed when two plates move towards each other, with one plate being forced over the top of the other plate. The highest mountains in the world are in the Himalayan range in Asia. These mountains have been created by the collision of two plates. The highest mountain in the world, Mount Everest, is growing by about 4 mm each year.

Around half of the heat in the mantle comes from radioactive decay of elements such as uranium and thorium. It is estimated that enough heat is released from the Earth's centre to heat 200 cups of coffee for each of the Earth's inhabitants per hour!

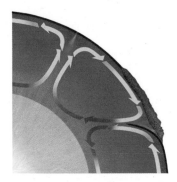

Figure 17.9 *Convection currents in the mantle of the Earth.*

Figure 17.10 *An Icelandic volcano released so much ash into the atmosphere in April 2010 that air travel was suspended in northern Europe for over a week.*

Volcanoes

Volcanoes occur at weak spots in the Earth's crust, where the crust is thin. Magma from the mantle rises through cracks which causes a build-up of pressure. When the tectonic plates of the Earth move, the crust can no longer contain the magma, forcing it out of the Earth in a volcanic eruption. Volcanic eruptions can be very spectacular, ejecting millions of tonnes of rock, magma and gas into the atmosphere.

? Did you know ...?

Magma that reaches the Earth's surface is called lava.

Figure 17.11 *Lava flowing from a Hawaiian volcano.*

N3 | L3 | N4 | L4

Activity 17.1

1. The Earth structure consists of layers of material at different temperatures. Identify the parts A, B, C and D in the diagram

2. The Earth's crust is made of three major types of rock:

(i) igneous, **(ii)** sedimentary and **(iii)** metamorphic.

For each type of rock:

(a) describe how it is formed **(b)** give two examples of that type of rock.

3. State the process that describes how rock is changed from one type to another.

4. The Earth's crust is composed of large areas of land called plates. The movement of these plates is called plate tectonics.

(a) What happens in the mantle of the Earth to cause the plates to move?

(b) State one of the major sources of heat in the mantle.

(c) Volcanoes often occur at tectonic plate boundaries. Describe the processes involved which cause a volcano to erupt.

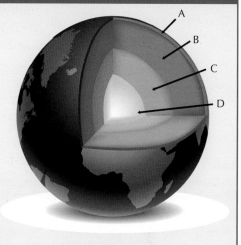

Weather and climate

Learning intentions

In this section you will:

- describe the difference between weather and climate
- identify factors that contribute to climate change
- learn the effect global warming is having on the planet.

National 3

Curriculum level 3

Planet Earth: Processes of the planet: SCN 3-05b

Weather and climate both describe different states of the atmosphere, but they relate to differing time scales.

Weather describes the atmospheric conditions over a short period of time, (daily or a few weeks). Analysis of weather and weather forecasts relates to effects on daily life and human activities over a relatively small geographical area.

Climate describes the average weather over a long period of time (years or decades). It describes the long-term weather patterns for a particular region or country.

Measuring the weather

The weather is the current condition of the air around us. Measurements can be made about different elements of the weather:

- temperature
- precipitation
- air pressure
- wind speed
- humidity
- cloud cover.

Table 17.2 *Measuring the weather*

Weather element	Measurement description	Measuring device
Temperature	How hot or cold the air is.	Thermometer
Precipitation	Any water in the air that falls under gravity (rain, sleet, snow, hail).	Rain gauge
Air pressure	The pressure exerted by the weight of air on the Earth.	Barometer
Wind speed	The average speed of the wind. Weather forecasts also have to take account of gusts, which are sudden increases in average wind speed that occur for short periods of time. Strong gusts can be very damaging.	Anemometer
Humidity	A measure of the amount of water vapour held in the air. Hotter air can hold more water vapour.	Hygrometer
Cloud cover	How cloudy the sky is.	The eye, and weather satellites

? Did you know ...?

An **okta** is the unit of measurement to determine the level of cloud cover in the sky.

Cloud cover					
Symbol	Scale in oktas (eighths)	Description	Symbol	Scale in oktas (eighths)	Description
 ○	0	Sky completely clear		 5	
 ◑	1			 6	
 ◔	2			 7	
 ◕	3			 8	Sky completely clear
 ◐	4	Sky half cloudy	⊗	 9	Sky obscured from view

GO! Activity 17.2

1. **Experiment** In this activity, you will measure the current weather conditions and create a short weather report.

Read the plan carefully then follow the instructions, making sure to follow all of the safety rules as you would normally be expected to do when carrying out practical work.

You can work in small groups but you should record your own observations and complete the answers yourself.

You will need: a thermometer, barometer, hygrometer (some electronic weather stations have all three), an empty 2-litre plastic bottle to make into a rain gauge, a measuring cylinder, scissors, some tape, a permanent marker, and an okta chart.

Many digital thermometers have built-in hygrometers for measuring humidity.

Method

(a) Make the rain gauge by cutting around the plastic bottle about two-thirds of the way up. Insert the top of the bottle upside down into the bottom of the bottle to act as a funnel. Fix it in place with tape. On another section of tape, draw a scale in cm with a permanent marker and attach it to the side of the bottle.

Ask your teacher to place the rain gauge in an wide open place the day before you take your weather measurements.

(b) If your school does not have an anemometer, you may wish to make one. Search online to find instructions on how to do that. Wind speed can then be recorded by the number of rotations of the anemometer in 10 seconds.

(c) Go outside and record the following:
   - temperature with the thermometer
   - air pressure with the barometer
   - humidity with the hygrometer
   - cloud cover referring to the okta scale
   - the amount of rain collected in the rain gauge
   - the wind speed, if you have an anemometer or have made one.

(d) Make a poster with your data, or record a short video clip to describe the weather conditions.

> ### Make the link
> Get help on this by reading the section of the Investigation Skills Appendix on presenting scientific findings.

Thunder and lightning

Thunderstorms are an example of an extreme form of weather. In a thunderstorm, thick clouds called cumulonimbus form when warm air quickly rises up through much colder air. This causes water vapour in the air to condense into water droplets.

> ### Make the link
> Warm air rises, because of convection currents. See Chapter 8 for more about convection.

N3	L3	N4	L4

As the warm air continues to rise, the water droplets cool to form ice crystals. Big enough ice crystals fall as hail to the bottom of the cloud. The hail gains a negative electric charge as it falls by rubbing against smaller, positively charged ice crystals. As the hail falls, the bottom of the cloud becomes negatively charged, and the top of the cloud becomes positively charged. This negative charge is attracted to the Earth's surface, and when the attraction is strong enough, the charges move to the ground quickly, creating a flash of lightning. The hail falls to the ground, often melting, to fall as heavy rain.

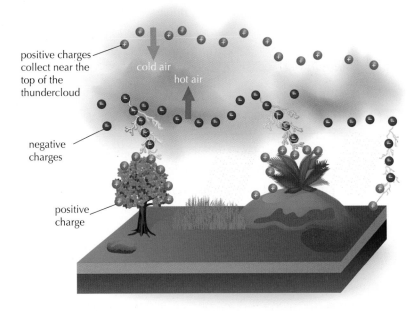

Figure 17.12 *Charges in storm clouds cause lightning.*

Make the link

The relationship between the temperature of a gas, such as air, and its volume is known as a gas law. It is explained in Chapter 8.

When the lightning passes through the air, it heats up the air, which expands rapidly. This pushes the air particles apart very quickly, creating vibrations. These vibrations are heard as thunder.

You can work out the distance to a flash of lightning by counting the time gap between seeing the flash of lightning and hearing the clap of thunder. The light of the lightning travels at the speed of light, 300 000 000 m/s, so the light only takes tiny fractions of a second to reach you. The sound of the thunder is much slower, travelling at the speed of sound in air, 340 m/s.

You can use your knowledge of the speed of sound to work out the distance:

Make the link

Find out more about the speed of sound in Chapter 10.

- in 1 second, the sound travels around 340 metres

- in 2 seconds, the sound travels 340 × 2 = 680 metres

- in 3 seconds the sound travels 340 × 3 = 1020 metres

and so on.

Example 17.1

The sound from a lightning bolt is heard 5 seconds after it was seen. How far away is it?

Time between seeing lightning and hearing thunder = 5 seconds

Distance sound travels in 1 second = 340 metres ●———(Write out what you know from the question.)

Distance to lightning bolt = distance the sound travels in 1 second × time ●——(Write out the equation.)

Distance to lightning bolt = 340 × 5 ●———(Substitute in what you know.)

$\quad\quad\quad\quad\quad$ = 1700 m ●———(Don't forget to write the units in your answer.)

GO! Activity 17.3

 1. **Experiment** In this activity, you will watch a demonstration of lightning using a Van de Graaff generator.

Read the plan carefully then follow the instructions, making sure to follow all of the safety rules as you would normally be expected to do when carrying out practical work.

This activity will be demonstrated by your teacher, but any observations can be recorded on your own.

You will need: a Van de Graaff generator.

Method

⚠ Sit a sensible distance away from the Van de Graaff generator when it is running – receiving a spark from one can be harmful.

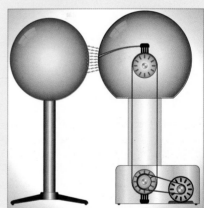

The belts in the Van de Graaff generator create the electric charge, which jumps to the other uncharged sphere.

 For best results this activity should be carried out on a dry day with low humidity. Turn on the Van de Graaff generator and allow sparks to jump between the belted sphere and the grounded sphere.

Draw a diagram of the Van de Graaff generator and compare its operation to how lightning is formed between clouds and the ground.

Our changing climate

Climate is the average weather that a region experiences over a period of around 30 years. The climate in Scotland varies from west to east. Temperatures are fairly similar throughout the year, but the west experiences more rain on average than the east, and usually has milder winters.

CLIMATE REGIONS

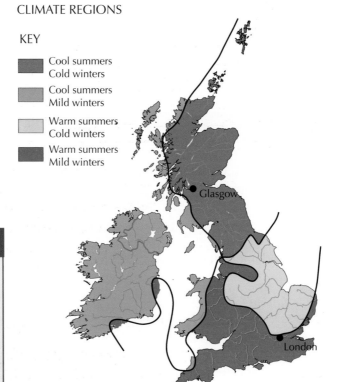

KEY

- Cool summers Cold winters
- Cool summers Mild winters
- Warm summers Cold winters
- Warm summers Mild winters

Figure 17.13 *Differences in the UK's climate.*

While weather can change many times over the course of a day, the climate can take many hundreds or thousands of years to change. Scientists are concerned that the climate is changing more rapidly than in the past.

Over the past 50 years, Scotland's climate has changed. The average temperature has increased by over 1 °C, with around half an hour more sunshine in the spring and autumn. Recent research suggests that Scotland is not significantly wetter than it used to be. However, it does rain more intensely than in the past. On average there are 8 more days of heavy rain than in the past.

? Did you know ...?

Record river levels in the river Don in north-east Scotland caused severe flooding in the winter of 2015/16. Thirty schools were closed during the flooding. December 2015 was the wettest month in the UK for over a century.

Figure 17.14 *Flooding in Ballater after Christmas in 2015 caused extensive damage to homes and businesses.*

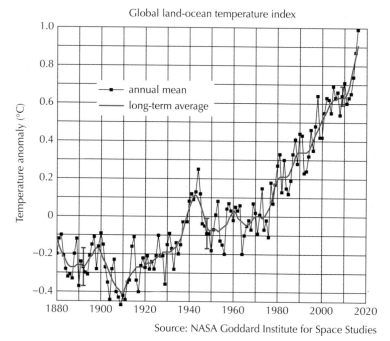

Figure 17.15 *Average temperatures on Earth have been on the rise for over a century.*

The graph in Figure 17.15 shows how the Earth's land and sea average temperature has changed over the past 140 years. Scientists call this temperature rise **global warming**.

Most of the warming over the last century has taken place because of the increased concentrations of greenhouse gases in the atmosphere.

Greenhouse gases trap the heat that the Earth radiates away, warming the atmosphere and the surface of the Earth. This is known as the **greenhouse effect**.

The most important greenhouse gases are carbon dioxide (CO_2), methane (CH_4), water vapour (H_2O) and nitrous oxide (N_2O).

📖 Word bank

• **Global warming**

Global warming is the term used by scientists to describe the observed rise in temperature of the Earth that has taken place over the course of the last century.

📖 Word bank

• **Greenhouse effect**

The greenhouse effect describes the warming of the Earth's atmosphere arising from the absorption of the Sun's energy by gases in the atmosphere.

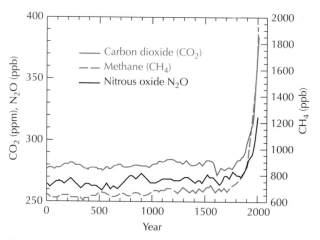

Figure 17.16 *Levels of greenhouse gases have greatly increased in the past 200–300 years.*

N3	L3	N4	L4

There are a number of reasons for the rises in the amounts of greenhouse gases in the atmosphere:

- increased burning of fossil fuels since the start of the Industrial Revolution in the second half of the eighteenth century for industry, electricity generation, heating and transport

- changes in agricultural practices causing increases in methane and nitrous oxide production

- deforestation, particularly in tropical rain forests, reducing the number of trees available to capture carbon

- rises in atmospheric temperature causing water vapour levels to rise

- increased world population, increasing the use of fossil fuels and changes in agricultural practices needed to feed billions more people.

The full effect of global warming on the planet is the subject of much debate. One likely outcome is that sea levels will rise because of the melting of glaciers and ice-caps. It is also thought that there will be a rise in extreme weather events such as storms and droughts.

Make the link

Renewable energy sources that do not produce greenhouse gases are described in Chapter 1.

Keep up to date!

Search online to find out what other consequences of global warming are likely to occur. Try to find out what governments and scientists are trying to do to slow the rate of global warming.

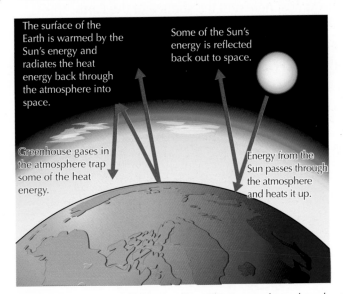

Figure 17.17 *The greenhouse effect. This image shows how heat is trapped by greenhouse gases in the atmosphere.*

GO! Activity 17.4

1. A student produced the table below to show how various elements of the weather can be measured, but did not finish it. Copy the table and complete it by adding the missing items.

Weather element	Measuring device
Temperature	
	Rain gauge
Air pressure	
	Anemometer
	Hygrometer
Cloud cover	

2. A family is camping in a thunderstorm and hears the sound of thunder 3 seconds after they see a flash of lightning. Given that the sound from the thunder travels 340 metres every second, how far away is the lightning?

3. A lady remarks one chilly afternoon, 'It is unusually cold for the time of year'. Explain how she has made an observation both about the weather and the climate.

4. Global warming is attributed to increasing greenhouse gases in the atmosphere.

(a) State two greenhouse gases that have increased in concentration in the atmosphere in recent years.

(b) Describe one effect of global warming on the planet.

Sun and Moon

National 3

Learning intentions

In this section you will:

- learn that day and night are caused by the Earth rotating on its axis
- learn that the Earth orbits the Sun once in one year
- learn how the seasons are caused by the tilt of the Earth
- learn how the rotation of the Moon around the Earth creates the phases of the Moon
- identify the phases of the Moon
- learn the effect the Moon has on the tides
- explain how a solar and lunar eclipse are formed.

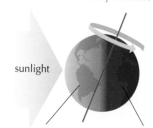

the Earth spins around its axis every 24 hours

sunlight

this side of the Earth is facing towards the Sun – it is day here

this side of the Earth is facing away from the Sun – it is night here

Figure 17.18 *Day and night are caused by the Earth spinning on its axis.*

The length of a day and a year is determined by the movement of the Earth and the Sun. A day is the time for the Earth to spin on its axis. A year is the time for the Earth to make one orbit of the Sun.

Day and night

As Figure 17.18 shows, it is daytime on the side of the Earth that is facing the Sun, and night-time on the opposite side of the Earth.

Activity 17.5

 1. Experiment In this activity, you will investigate how the position of the Sun varies throughout the day, and how this affects the length of the shadow formed by a person.

Read the plan carefully then follow the instructions, making sure to follow all of the safety rules as you would normally be expected to do when carrying out practical work.

This activity will be demonstrated by your teacher, but any observations can be recorded on your own.

You will need: a model Earth or globe, a lamp, a short stick figure, and some Blu-tak.

Method

Set up the globe and lamp in a dark room as shown in the picture.

(a) Attach the stick figure to the globe with Blu-tak, so that it is standing on the Northern Hemisphere, in Europe. The lamp represents the Sun. With the figure positioned in darkness, rotate the globe anti-clockwise so that the figure just moves into daylight. This is morning. What direction on the Earth does the figure have to look to see the Sun?

(Hint: is it looking north, south, east or west?)

What is the length of the figure's shadow?

(b) Continue rotating the globe so that the figure is now directly in front of the Sun. This is midday. Notice that from its position in the Northern Hemisphere, to look at the Sun the figure must face directly south. How does this compare with the direction we must look to see the Sun in the middle of the day? What is the length of the figure's shadow now?

> ## 📖 Word bank
>
> • **Hemisphere**
>
> The word 'hemisphere' means 'half of a sphere'. The Equator splits the Earth into the Northern Hemisphere and the Southern Hemisphere.

(c) Continue rotating the globe so that the figure is now almost in the darkness again. This is evening. What direction must the figure face now to look at the Sun? What is the length of its shadow?

(d) Copy and complete the table below to show what you have learned from this demonstration.

Time of day	Direction to look to see the Sun	Length of shadow
Morning		
Midday		
Evening		

The seasons

The Earth's axis of rotation is tilted at 23°. In the winter, the Northern Hemisphere is tilted away from the Sun. The Sun is above the horizon for shorter times each day, so there is less daylight. The Sun is low in the sky and the amount of heat from the Sun is reduced.

Six months later, in the summer, the Earth is on the other side of the Sun and the Northern Hemisphere is tilted towards the Sun. The Sun is high in the sky, there is more daylight and the heat from the Sun is increased.

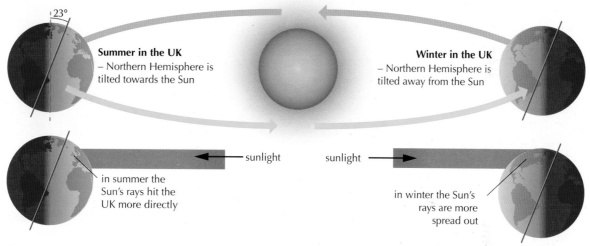

Figure 17.19 *The seasons are caused by the Earth's tilt.*

Regions close to the North Pole above the Arctic Circle experience 24 hours of daylight in the summer time, and 24 hours of darkness in the winter time.

The Moon

The Moon is a natural satellite that orbits the Earth in 29 days. This period is called a lunar month. The Moon is the second brightest object in the sky (after the Sun) but it does not emit any of its own light. Instead, it reflects the light from the Sun.

The Moon's appearance changes at different times of the lunar month. These changes in shape are known as the 'phases' of the Moon.

> **? Did you know ...?**
>
> The axis of the planet Uranus is tilted at 82°. This means it rotates on its side as it orbits the Sun.

> **Make the link**
>
> Natural and artificial satellites are explored in Chapter 16.

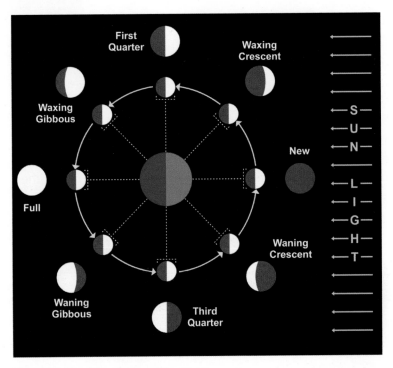

Figure 17.20 *The phases of the Moon.*

Regardless of its position, one half of the Moon is always illuminated by the Sun. The way we see it depends on how much of the illuminated half we can see. When we see all the illuminated half, we see a complete disc, known as a full Moon. When we only see part of the illuminated half, we see a **waning** or **waxing** crescent Moon. The changing position of the Moon as it orbits the Earth allows us to see the different phases.

Figure 17.21 *A full Moon above Edinburgh castle.*

📖 Word bank

• **Waning**

Waning means the phase of the Moon is shrinking. A waning gibbous Moon describes what we see when more than half of the Moon is illuminated and is getting less bright.

• **Waxing**

Waxing means the phase of the Moon is growing. A waxing gibbous Moon describes what we see when more than half of the Moon is illuminated and is getting brighter.

Activity 17.6

 1. Experiment In this activity you will investigate the phases of the Moon.

Read the plan carefully then follow the instructions, making sure to follow all of the safety rules as you would normally be expected to do when carrying out practical work.

This activity will be demonstrated by your teacher, but any observations can be recorded on your own.

You will need: an office (swivel) chair, a lamp, a white polystyrene ball and a pencil or stiff wire.

Method

(a) The lamp represents the Sun. The Moon is represented by the white ball and you are the Earth.

Using a polystyrene ball to model the phases of the Moon.

(b) Push the pencil/stiff wire into the polystyrene ball so that the ball can be held upright with the pencil. Sit in the chair and hold the ball at arms' length in front of you (and slightly above your head), facing the lamp. The side of the Moon facing the lamp will be white. The side facing you will be dark. What is the phase of the Moon in this position?

(c) Rotate anti-clockwise on the chair a quarter turn. Keep holding the Moon directly in front of you. What is the phase of the Moon in this position? What phase of the Moon did you see during this first quarter turn?

(Hint: use the diagram in Figure 17.20 to help you.)

(d) Rotate in your chair until the lamp (the Sun) is directly behind you. Look at the Moon (ensure that it is above your head so the light from the lamp is not blocked from hitting the Moon). What phase is it now? What phase did you pass through during this quarter turn?

(e) Rotate in your chair anti-clockwise another quarter turn. Look at the Moon. What phase is now visible and what phase did you see during the quarter turn?

(f) Complete one last quarter turn anti-clockwise. You should be facing the Sun again. What was the last phase of the Moon that you observed before returning to the new Moon?

(g) If you have a video camera, record a short video of the phases of the Moon during one complete anti-clockwise rotation. Add a voice-over to describe the phases of the Moon as you rotate.

Solar and lunar eclipses

From the Earth, the Sun and Moon look like they are the same size. In fact, the Sun is about 400 times the diameter of the Moon. It looks a similar size because it is also around 400 times further away from the Earth. As they have the same apparent size, the disc of the Moon is the right size to block out the light from the Sun when the Moon's orbit takes it directly between the Sun and the Earth. This is known as a **solar eclipse**.

Figure 17.23 *The last solar eclipse in Scotland was a partial eclipse on 20 March 2015.*

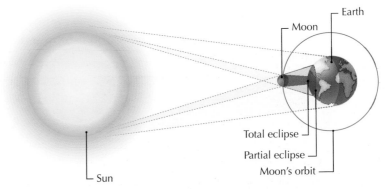

Figure 17.22 *In a solar eclipse, the Sun's light is blocked by the Moon.*

In a lunar eclipse, the Moon passes directly behind the Earth into its shadow. This happens when the Earth is directly between the Sun and the Moon.

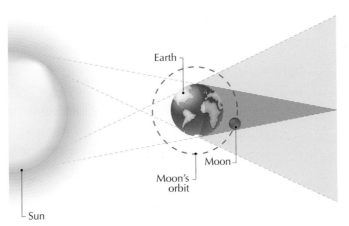

Figure 17.24 *In a lunar eclipse, the Earth blocks the Sun's light to the Moon.*

Activity 17.7

 1. Experiment In this activity, you will observe a model demonstrating solar and lunar eclipses.

Read the plan carefully then follow the instructions, making sure to follow all of the safety rules as you would normally be expected to do when carrying out practical work.

This activity will be demonstrated by your teacher, but any observations can be recorded on your own.

You will need: a lamp, a large ball or globe (representing the Earth) and a ping pong ball (representing the Moon). Two different-sized polystyrene balls connected with stiff wire would also work, as shown in the figure here.

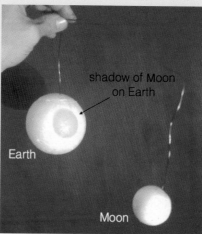

The shadow of the ping pong ball on the Earth is a solar eclipse.

 Method

(a) Turn on the lamp in a darkened room and position the ping pong ball between the lamp and the large ball. Draw a diagram of what you see and explain whether it represents a solar or lunar eclipse.

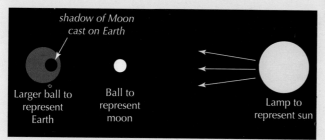

Arrangement of a lamp and small and large balls to model a solar eclipse.

(b) Now place the large ball between the lamp and the ping pong ball (representing the arrangement in Figure 17.24), so that the large ball casts a shadow on the ping pong ball. Draw a diagram of what you see and state whether it is a solar eclipse or lunar eclipse.

Tides

The phases of the Moon have a significant impact on the Earth's tides. As the Earth spins on its axis, the pull of gravity from the Moon causes the oceans to bulge out in the direction of the Moon. The Sun's gravity also has an impact on the tides.

When the Sun and Moon are aligned, the tidal bulges are greater than normal, so the tides are higher than normal. The resulting **neap tides** take place during a new Moon or a full Moon.

When the Sun and Moon are not aligned, the high tides are lower, and are called **spring tides**.

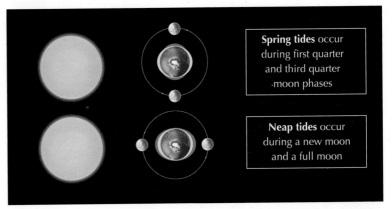

Spring tides occur during first quarter and third quarter moon phases

Neap tides occur during a new moon and a full moon

Figure 17.25 *Spring and neap tides – high tides vary in height throughout the month due to the phase of the Moon.*

There are usually two high tides each day because the Earth rotates through two tidal bulges in 24 hours.

GO! Activity 17.8

1. Explain why the Earth experiences day and night.

2. State how long the Earth takes to orbit the Sun.

3. The figure below shows how the Northern Hemisphere tilts towards the Sun in the summer. By referring to the diagram, explain why the Sun never sets at the North Pole in the summer.

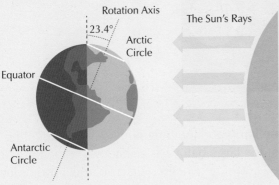

4. A child notices that the shadow of a lamppost changes throughout the day.

   (a) State two ways in which the shadow changes throughout the day.

   (b) For each of your answers to part (a), explain why the change takes place.

5. Plants need light and heat to grow. Explain why they grow better in summer than they do in winter.

6. The Moon does not emit any light. Explain why we can still see the Moon.

7. Copy and complete the following paragraph using the words provided.

   Earth   **phases**   **lunar**   **29**   **shape**

   The Moon takes _ _ days to orbit the _ _ _ _ _ . This is called a _ _ _ _ _ month. During a month, the _ _ _ _ _ of the Moon changes, having different _ _ _ _ _ _.

8. The Moon is directly in between the Earth and the Sun during an eclipse.
   (a) What is the phase of the Moon at this moment?
   (b) What type of eclipse is taking place?
   (c) What are the tides like at this moment?

The Solar System

National 3

Learning intentions

In this section you will:

- state that the Solar System is made up of planets and other objects that orbit the Sun
- learn about the size of the planets relative to each other
- learn about the distances between each planet and the Sun
- learn about comets and other small objects that orbit the Sun
- state the difference between a meteor and a meteorite.

The Earth is one of eight planets that orbit the Sun. Together with other objects like comets, asteroids and dwarf planets, they make up the Solar System.

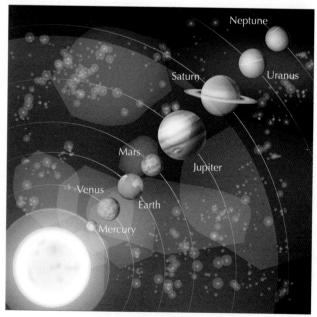

Figure 17.26 *The planets of the Solar System orbit the Sun. (Distances and planet sizes are not shown to scale.)*

📖 Word bank

- **Dwarf planet**

An object in the Solar System that does not meet all the criteria to be classed as a normal planet is known as a dwarf planet.

★ Memory Aid

There are a number of mnemonics to help you remember the order of the planets in the Solar System. Here is one:

- My Very Excellent Mother Just Served Us Noodles.

Can you think of any others?

The eight planets, in order of closest to the Sun first, are:

Mercury, Venus, Earth, Mars, Jupiter, Saturn, Uranus and Neptune.

The planets are not the only objects orbiting the Sun. There are also thousands of **asteroids** orbiting the Sun, mainly in a belt between Mars and Jupiter. Asteroids are lumps of rock smaller than planets.

Comets are mainly found beyond Neptune, at the outer edge of the Solar System. Comets are made of ice and rock. When a comet's orbit brings it close to the Sun, some of the ice turns into gas, creating a visible tail. There are about 5000 known comets orbiting the Sun.

? Did you know ...?

The Rosetta space probe was sent on a 10-year mission in 2004 to catch and land on a comet. It landed successfully on a comet called 67P, in November 2014.

Figure 17.27 *Comet 67P photographed by the Rosetta probe, two months before the probe landed on it.*

Figure 17.28 *Comet Lovejoy, taken by a NASA astronaut on the International Space Station.*

Meteors are bits of rock or dust that have broken off asteroids or comets. They get trapped by the pull of gravity of the Earth, and burn up as they pass through the atmosphere at speeds of up to 72 km/s.

Meteors are often called shooting stars, as they leave a trail of gas and melted particles when they burn up and appear in the night sky as streaks of light. Meteors that do not fully disintegrate before reaching the surface of the Earth are called **meteorites**.

↻ Keep up to date!

Meteor showers happen when a lot of meteors fall through the night sky at the same time. They occur fairly regularly over the course of a year. Search online to find out when the next visible meteor shower is taking place. If it is a clear night, why not go outside and try to spot one?

GO! Activity 17.9

☻ The table below contains some data on the eight planets in the Solar System. The inner planets exist between the Sun and the asteroid belt. The outer planets exist beyond the asteroid belt.

	Planet	Average distance from the Sun (millions km)	Diameter (km)
Inner planets	Mercury	58	5000
	Venus	108	12 000
	Earth	150	12 800
	Mars	228	6800
Outer planets	Jupiter	780	143 000
	Saturn	1400	121 000
	Uranus	2900	51 000
	Neptune	4500	50 000

1. Use the table to answer the following questions.
   (a) How many times further away from the Sun is Neptune compared to the Earth?
      (Hint: Divide the average distance Neptune is to the Sun by the average distance the Earth is to the Sun.)
   (b) How does the size of the four inner planets compare to the size of the four outer planets?
   (c) Search online to research why the size of the four outer planets is so different from the four inner planets.
2. Three boys are watching the stars while out camping. One boy spots a shooting star and thinks it is a comet. His friend thinks it is a meteor, and the third says it is a meteorite. Explain the differences between comets, meteors and meteorites to explain what the boys might have seen in the sky.

GO! Activity 17.10

☻ 1. **Experiment** In this activity you will create a scale model of the Solar System and Pluto on a 1 m strip of paper.

Read the plan carefully then follow the instructions.

You will need: a 1-metre long strip of paper and a pencil or pen.

Method
   (a) Label one end of the paper **Sun** and the other end of the paper **Pluto**.
   (b) Fold the paper in half and crease it. Open it up and mark the halfway point as **Uranus**.
   (c) Fold the paper again in half and half again, then unfold it flat. The paper strip is now divided into quarters. Label the first quarter, closest to the Sun, **Saturn** and the third quarter **Neptune**.

(continued)

(d) There are now five planets to fit in the first quarter of the paper strip, between the Sun and Saturn.

(e) Fold the first quarter of the paper (between the Sun and Saturn) in half, crease and unfold it. Mark **Jupiter** at this new crease (1/8th of the whole strip).

(f) Now fold the Sun out to meet Jupiter, crease it and unfold it. This new crease (1/16th of the whole strip) can be marked **Asteroid Belt**.

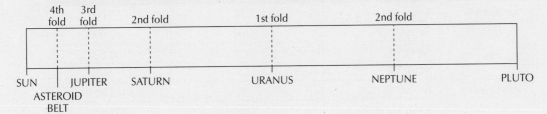

How the paper should look by the end of step 6.

(g) Folding now gets a bit more difficult! Fold the Sun out to meet the Asteroid Belt, crease and unfold. This new crease can be labelled **Mars**.

(h) Three more planets to go – Mercury, Venus and Earth. Fold the Sun to meet Mars, then fold that section again. Opening it should reveal three creases. Mark on these creases **Mercury**, **Venus** and **Earth**, with Mercury closest to the Sun.

(i) Open the entire strip and admire your work!

(j) Here are three interesting facts related to your model:

- there is a lot of empty space in the Solar System
- at this scale, the Sun would be the size of a grain of sand, and the planets would not be visible to the eye
- at this scale, the nearest star to the Sun would be **4.2 miles** away.

Learning checklist

After reading this chapter and completing the activities, I can:

N3 L3 N4 L4

- identify the different sections of the structure of the Earth. **Activity 17.1 Q1** ○ ○ ○

- state that the three major types of rock in the Earth's crust are igneous, sedimentary and metamorphic, and provide examples of each. **Activity 17.1 Q2** ○ ○ ○

- state that the rock cycle describes the processes of changing rock from one type to another. **Activity 17.1 Q3** ○ ○ ○

- state that convection currents in the mantle of the Earth cause plates in the Earth's crust to move, causing earthquakes. **Activity 17.1 Q4** ○ ○ ○

N3 L3 N4 L4

- describe how volcanoes occur. **Activity 17.1 Q4**

- give examples of devices used to measure different types of weather. **Activity 17.2, Activity 17.4 Q1**

- state the difference between weather and climate. **Activity 17.4 Q3**

- identify factors that contribute to climate change. **Activity 17.4 Q4**

- state the effect global warming is having on the planet. **Activity 17.4 Q4**

- state that day and night are caused by the Earth rotating on its axis. **Activity 17.5, Activity 17.8 Q1**

- state that the Earth orbits the Sun once in one year. **Activity 17.8 Q2**

- explain how the seasons are caused by the tilt of the Earth. **Activity 17.8 Q3, Q5**

- explain how the rotation of the Moon around the Earth creates the phases of the Moon. **Activity 17.6, Activity 17.8 Q7**

- identify the phases of the Moon. **Activity 17.6, Activity 17.8 Q7**

- state that the Moon causes high tides in the Earth's oceans. **Activity 17.8 Q8**

- explain how solar and lunar eclipses are formed. **Activity 17.7, Activity 17.8 Q8**

- state that the Solar System is made up of planets and other objects that orbit the Sun. **Activity 17.9 Q1, Activity 17.10**

- state the sizes of the planets relative to each other. **Activity 17.9 Q1**

- represent the distances between each planet and the Sun using a scale diagram. **Activity 17.10**

- explain the difference between a comet, meteor and meteorite. **Activity 17.9 Q2**

18 Cosmology

This chapter includes coverage of:

N4 Cosmology • Planet Earth SCN 3-06a • Planet Earth SCN 4-06a

You should already know:

- how the length of a day, a month and a year relate to the patterns of movement of the Sun and moon
- that the Solar System consists of eight planets that orbit the Sun
- the relative size and scale of the planets in the Solar System.

National 4

Cosmology

Learning intentions

In this section you will:
- state and explain what is meant by the following terms: planet, moon, star, Solar System, exo-planet, galaxy and universe
- learn how to present solar system data graphically
- define a light year as the distance light travels in one year
- convert distances in metres into light years.

Cosmology is the scientific study of how the universe has changed over time. Our understanding of the universe, and our place in it, has developed over centuries through observation and exploration. Astronomers have gained greater understanding of many aspects of the universe. Descriptions of some important astronomical objects are given in Table 18.1.

| N3 | L3 | **N4** | L4 |

Table 18.1 *Important astronomical objects*

Astronomical object	Description
Planet	An object big enough to have its own gravitational field that orbits a star. The Earth is a planet.
Moon	A natural satellite that orbits a planet. The Moon orbits the Earth. It does not create its own light, but instead reflects the Sun's light.
Star	A huge sphere of gas that emits light and heat. Our Sun is the nearest star to the Earth.
Solar System	A star and all the objects that orbit around it, including planets, moons, asteroids and comets. The Sun is at the centre of our Solar System, and eight planets orbit it.
Exo-planet	A planet that orbits a star outside our Solar System.
Galaxy	A huge collection of stars, dust and gas, held together by gravity. Our Sun is in the Milky Way galaxy.
Universe	All of matter, space and time.

The universe is incomprehensibly huge. Planet Earth is a relatively small planet that orbits an average sized star, the Sun.

> **Make the link**
>
> Satellites are explained in more detail in Chapter 16.

Figure 18.1 *The planets in the Solar System to scale, next to the Sun. (Note the distances between the planets and the Sun are not to scale.)*

The Sun is one of around 100 billion stars that make up our galaxy, called the Milky Way. Many of these stars may have planets, known as exo-planets, which orbit them. The Milky Way galaxy is just one of millions of galaxies that make up the universe.

> **? Did you know ...?**
>
> The Milky Way galaxy was originally named by the Romans. They named it Via Lactea, which translated means *The Road of Milk*. They gave it this name because it looks like milk splashed across the sky.
>
> Since the Solar System is inside the disk of the Milky Way, this is how we see our galaxy, when looking at it from one of its spiral arms.
>
>
>
> **Figure 18.2** *The Milky Way is visible on a clear night.*

GO! Activity 18.1

☺ In this activity, you will analyse data about the Solar System to find relationships.

The table contains information about the eight planets in our Solar System.

Planet	Mercury	Venus	Earth	Mars	Jupiter	Saturn	Uranus	Neptune
Diameter (km)	5000	12000	12800	6800	143000	121000	51000	50000
Distance from Sun (million km)	58	108	150	228	780	1400	2900	4500
Period of orbit around Sun (days)	88	225	365	687	4330	10800	30600	59800
Mass of planet ($\times 10^{24}$ kg)	0.33	4.9	6.0	0.64	1900	570	87	100
Strength of gravity ($N\,kg^{-1}$)	3.7	8.9	9.8	3.7	23	9.0	8.7	11
Average surface temperature (°C)	167	464	15	−65	−110	−140	−195	−200
Number of moons	0	0	1	2	67	62	27	14

1. Plot a bar graph to show how the number of moons varies for each planet.

2. Search online to find out why some planets have no moons and others have many moons.

3. Plot a line graph to show how the temperature of the planets varies with distance from the Sun.

4. Describe the shape of the graph you have drawn for question 3.

5. The dwarf planet Pluto is further from the Sun than Neptune. Suggest a value for the average surface temperature of Pluto.

6. The Earth takes one year to orbit the Sun. How many years does it take Neptune to orbit the Sun?

7. Use the data in the table to answer these questions about the mass of the Sun and the planets.

   (a) How many times heavier is Jupiter than the Earth?

   (b) What is the total mass of all the planets in the Solar System?

   (c) The mass of the Sun is 2×10^{30} kg. How many times heavier is the Sun than all the planets combined?

🔍 Hint

The scales on both axes will be tricky for this graph. The horizontal axis (distance from Sun) will range from 0 to 4500, and many of the planets will be very close to the zero end. The vertical axis (temperature) will have both positive and negative values.

🔍 Hint

How many times bigger is the Sun?

8. The Sun has a diameter of 1.39 million km. Compare this with the diameter of the Earth.

9. Look at the properties of the planets listed in the table. What other properties do you think might be related to each other? How would you test your idea?

The light year

The distances between the stars in our galaxy are enormous. For example, Proxima Centauri, the nearest star to the Sun, is 40 000 000 000 000 km away (40 trillion kilometres). With such enormous numbers, it is not practical to use metres or kilometres as the unit of distance. Instead of using kilometres to measure astronomical distances, the **light year** is used. One light year is equal to the distance light can travel in one year, in metres.

To convert one light year to metres, consider the distance light can travel in one year.

distance = speed × time

In this equation,

speed = the speed of light ($300\,000\,000\,ms^{-1}$)

time = number of seconds in one year

1 year = 365 days

$\qquad$ = (365 × 24) hours

$\qquad$ = (365 × 24 × 60) minutes

$\qquad$ = (365 × 24 × 60 × 60) seconds = 31 536 000 s

distance = 300 000 000 × 31 536 000

distance = $9.5 \times 10^{15}\,m$

The distance travelled by light in one year is $9.5 \times 10^{15}\,m$.

This is an incredibly large distance. It is equivalent to travelling 238 million times round the Equator, or over 12 million return trips to the Moon.

Using light years, we can gain a better appreciation for how big the universe is.

- It takes 8 minutes for light to travel from the Sun to the Earth so the distance from the Sun to the Earth is 8 light minutes.

📖 Word bank

- **Light year**
A light year is the distance travelled by light in one year.

❓ Did you know …?

One light year is a huge distance. Travelling at the speed of the NASA New Horizons spacecraft that reached Pluto in 2015, it would take around 78 000 years to travel to Proxima Centauri.

- The distance from the Sun to Pluto is 327 light minutes.
- The distance to the nearest star is 4.2 light years.
- The Milky Way galaxy is 100 000 light years in diameter.
- The distance to the Andromeda galaxy (our nearest galaxy) is 2.5 million light years.
- It is estimated that the universe is 93 billion light years in diameter.

Example 18.1

Proxima Centauri is 4.0×10^{16} m away. Determine the distance to Proxima Centauri in light years.

Distance to Proxima Centauri = 4.0×10^{16} m

1 light year = 9.5×10^{15} m •————————————⟨ Write out what you know from the question. ⟩

Distance to Proxima Centauri in light years

$$= \frac{\text{distance to star in metres}}{\text{distance of 1 light year in metres}} = \frac{4.0 \times 10^{16}}{9.5 \times 10^{15}}$$

Distance to Proxima Centauri = 4.2 light years

GO! Activity 18.2

☺ 1. Explain the difference between each of the following astronomical objects:
   (a) planet and moon
   (b) star and Solar System
   (c) galaxy and universe.

2. A student makes the following statement about the Sun and the Moon.
   'The Sun provides light for the day and the Moon provides light for the night.'
   Explain the difference between how the Sun and the Moon provide light.

3. The Sun has eight planets that orbit it. What is the name given to planets that orbit other stars?

4. Two students are having an argument about what a light year is. The first student thinks it is a measure of distance. The second student believes it is a measure of time. Describe what a light year is and state which student is correct.

5. In the Northern Hemisphere, the North Star is the point around which the northern sky appears to turn. It can be used for navigation, since it will always show the direction of due North.

N3　L3　**N4**　L4

The North Star is around 433 light years away.

(a) How many years does it take light to travel from the North Star to the Earth?

(b) Use the equation *distance = speed × time* to show that 1 light year is equal to 9.5×10^{15} m.

(c) Calculate the distance to the North Star in metres.

Space observation and exploration

National 4
Curriculum level 4
Planet Earth: Space
SCN 4-06a

Learning intentions

In this section you will:

- give examples of methods used to observe and explore space
- give examples of the impact that space observation and exploration has had on our understanding of the universe and planet Earth.

Until the middle of the twentieth century, the only way to explore space was through observing it by eye. People have been observing the sky for centuries, and there are many examples of star catalogues and atlases, dating back thousands of years. Galileo was an Italian astronomer who first used a telescope to look at the sky. In January 1610 he used it to discover three of the Moons of Jupiter. Galileo was one of the first astronomers to suggest that the Earth orbited around the Sun.

Between Galileo's day and the 1960s, many ground-based observatories were built to observe the light coming from space using telescopes. Scotland has a number of working observatories. The most recent is the Scottish Dark Sky Observatory in the Galloway Forest Park. The lack of **light pollution** in that area makes it ideal for observing the night sky.

Figure 18.3 *Galileo is often called the father of modern astronomy.*

📖 Word bank

- **Light pollution**

The night sky in towns and cities is often brightened by street lights and other artificial sources. This is called light pollution; it makes observation of the stars and planets more difficult in built-up areas.

? Did you know ...?

Scotland has led the way on a number of recent discoveries in space. In 2015, scientists at Edinburgh University discovered a planet that floats in space without a star to orbit. It has clouds made of molten iron.

In 2016, researchers at Glasgow University lead a team that confirmed the existence of gravitational waves. These waves were originally predicted by Albert Einstein in 1916, 100 years earlier. They used detectors in America that can measure vibrations that are a thousand-millionth the size of a single wavelength of light!

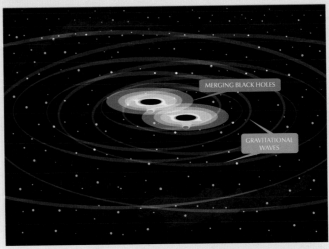

MERGING BLACK HOLES

GRAVITATIONAL WAVES

Figure 18.4 *Gravitational waves are created by some of the most violent events in the universe, like the collision of two black holes.*

⟳ Keep up to date!

The James Webb Space telescope is a large infrared telescope currently in development. It is planned to be launched in October 2018. It aims to support the study of the history of the Universe, and the formation of solar systems that would be capable of supporting life on planets like Earth.

Search online to research how this telescope has been designed and developed to study the farthest reaches of the universe.

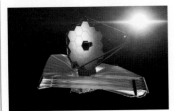

Figure 18.5 *The James Webb telescope*

In 1968, America launched the first telescope into orbit. These telescopes (also known as space observatories) eliminate the problems created by light pollution and the distortion of light due to the atmosphere. Space observatories, such as the Hubble Space Telescope and the Kepler observatory, have significantly improved astronomers' understanding of the universe.

The Hubble Space Telescope can view light from all parts of the electromagnetic spectrum. Images from the Hubble Space Telescope have helped astronomers understand how the universe has changed over time.

The Kepler observatory was launched in 2009 and is designed to discover Earth-sized planets orbiting round other stars. It does this by detecting tiny variations in the light from stars as planets pass in front of them. In early 2017, NASA confirmed that the Kepler observatory had detected 2331 planets.

☀ Physics in action: Extreme observations

Stars and other stellar objects in space emit light over the full range of wavelengths in the electromagnetic spectrum, not just visible light. By observing the light from different parts of the spectrum, our understanding of the universe can increase.

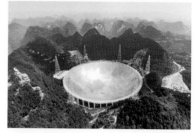

Figure 18.6 *China's radio telescope as seen from the air.*

Radio telescopes are used to observe the radio waves emitted from stars. Since radio waves have such long wavelengths, radio telescopes also need to be large. China completed construction of the world's largest radio telescope in July 2016. Known as 'China's Eye of Heaven', it is as large as 30 football pitches. It is being used to detect radio waves from faint rotating stars called pulsars, as well as search for signs of extra-terrestrial communication and intelligence.

Radio telescopes are also used to study exploding stars. By using the radio wave data, astronomers can reconstruct the physics of stellar explosions.

Figure 18.7 *The shockwaves from exploding star Supernova 1987A are studied using radio telescopes.*

Visible light emitted from stars can be analysed to identify elements present in the stars. Each element produces its own unique line spectrum. By comparing the light from stars with the line spectra from different elements, astronomers can discover what stars are made from. No two elements have the same line spectra, so they can be used like fingerprints to identify different elements and combinations of elements.

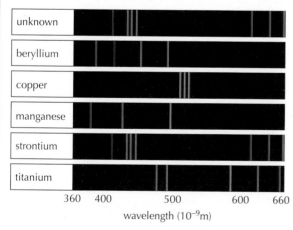

Figure 18.8 *From this chart you can identify which element the unknown line spectrum is from. The unknown line spectrum comes from strontium.*

Space exploration

Early space exploration started with unmanned satellites launched by the USSR (now Russia) and the USA. The first man in space was Yuri Gagarin in 1961. His Vostok 1 spacecraft completed one orbit of the Earth.

Throughout the 1960s, the American space program had the aim of landing on the Moon. Neil Armstrong and Buzz Aldrin were the first astronauts to land on the Moon on the Apollo 11 mission in 1969.

By 2016, approximately 550 people have travelled into space. Of those, most of them have been either American or Russian. Only two have come from the UK; Helen Sharman and Tim Peake.

Table 18.2 *Significant milestones in space exploration.*

Date	Event	Mission	Nation
October 1957	First artificial satellite – Sputnik 1	Sputnik 1	USSR
April 1961	First human spaceflight – Yuri Gagarin	Vostok 1	USSR
June 1963	First woman in space – Valentina Tereshkova	Vostok 6	USSR
February 1966	First soft landing on the Moon	Luna 9	USSR
July 1969	First human on the Moon	Apollo 11	USA
December 1971	First soft landing on Mars	Mars 3	USSR
April 1981	First reusable manned spaceflight	STS-1	USA
August 1989	First Neptune fly-by	Voyager 2	USA
February 1990	Optical orbiting observatory launched	Hubble Space Telescope	USA and Europe
July 1997	First rover on another planet (Mars)	Mars Pathfinder	USA
November 1998	First multinational space station	International Space Station	Russia, USA, Europe, Japan, Canada
March 2009	Kepler Mission launched, to search for Earth-like planets around other stars	Kepler Mission	USA
July 2015	First fly-by of Pluto	New Horizons	USA

↻ Keep up to date!

The list of significant milestones in Table 18.2 is only a snapshot of some major firsts in space exploration. Why not create a timeline of your own, focusing on a specific type of exploration. Pick from the list below and search online to research the major events related to it.

- Moon landings

- Probe fly-bys of other planets

- Telescope launches (e.g. Hubble or Kepler)

- Satellite launches

- Reusable manned spacecraft (e.g. the Space Shuttle)

- Exploration of Mars (e.g. Pathfinder, Mars rovers Spirit, Opportunity and Curiosity)

Figure 18.9 *In total, only 12 people have ever walked on the Moon.*

In 1969, Neil Armstrong and Buzz Aldrin landed on the Moon and undertook the first spacewalks. Since then, only another 10 American astronauts have walked on the Moon. More recently, most astronauts have used the International Space Station as a base from which to explore life in space.

Space probes are used to explore the Solar System. Many nations have sent probes into space to explore other planets or moons, as well as orbit asteroids or comets. Missions involving space probes usually last for many years, due to the enormous distances involved in travelling to planets within the Solar System.

Probes have successfully been sent to all the planets in the Solar System. The Voyager 1 probe is the longest-lasting NASA mission to date. It was launched on 5th September, 1977, and has explored Jupiter and Saturn and their moons.

Having successfully visited Jupiter and Saturn, it has since travelled out of the Solar System into interstellar space (the space between the stars). It is still operating and sending data back to Earth. In early 2017, it was approximately 135 **astronomical units** (AU) from the Sun and is expected to remain operational until 2025.

Some probes land robots on planets. The robotic rover Curiosity has been on Mars since 2012 and has studied its climate and geology.

📖 Word bank

- **Space probe**

A space probe is a robotic, unmanned spacecraft that is sent from Earth to explore space.

Figure 18.10 *The Great Red Spot on Jupiter taken by Voyager 1 in February 1979.*

📖 Word bank

- **Astronomical units**

Astronomical units (AU) are a convenient unit for measuring large distances in space. The distance between the Earth and the Sun is 1 AU.

Figure 18.11 *The Mars rover Curiosity.*

⟳ Keep up to date!

As well as the Mars probe mission, there are many other probes currently operating around other planets.

Akatsuki is the first Japanese probe to visit Venus. It is studying the atmosphere of the planet after obtaining orbit in December 2015.

China's Lunar Exploration Program has launched a series of probes to study the Moon. These missions, starting in 2007, have included orbiting probes, landing rovers and sample return spacecraft.

The European Space Agency (ESA) currently has a probe orbiting Mars, called the Exomars Trace Gas Orbitor (TGO). It is searching for gases in the Martian atmosphere that could provide evidence for life.

Search online to find out the current status of these or other probe missions.

GO! Activity 18.3

☺ 1. Give three examples of methods used to observe and explore space.

2. Observatories on Earth and in space use telescopes to make observations of space. State two advantages of using telescopes to observe space over simply using our eyes.

3. Space probes have allowed astronomers to explore the far reaches of the Solar System. Search online to find out five things that we would not know about space without space probes.

Curriculum level 3

Planet Earth: Space

SCN 3-06a

National 4

📖 Word bank

- **Habitable zone**

The region of space around a star that a planet or moon must exist to sustain life is called the habitable zone. A planet in this region will have a temperature at which water can exist in liquid form.

Life outside the Solar System

Learning intentions

In this section you will:

- identify the conditions required for an exo-planet to sustain life
- produce a reasoned argument for the likelihood of life existing elsewhere in the universe.

Currently, we know of only one planet in the universe that contains life – planet Earth. Astronomers believe that in order for a planet to sustain life, it must exist within the **habitable zone** of a star.

This region of space around the Sun is also known as the 'Goldilocks' zone. Earth is in the Goldilocks zone – it is not too hot, and not too cold, but just the right temperature for liquid water to exist. Within the Solar System, Mars and the Moon also exist within the habitable zone of the Sun. Mars is generally very cold, with an average temperature of −67°C.

However, for short periods of time some regions of Mars have temperatures and air pressure that would allow water to exist in liquid form. As yet there is no direct evidence of liquid water being found on Mars.

For a planet to be habitable, its environment must provide organisms with a dependable supply of the basic needs for life. As well as liquid water, other requirements would include:

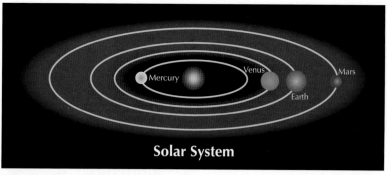

Figure 18.12 *The habitable zone around the Sun. Venus orbits just on the edge of the habitable zone.*

- nutrients for vegetation to grow
- a breathable atmosphere (one which contains oxygen)
- an energy supply such as a nearby star (to provide both light and heat)

Exo-planets

Exo-planets are planets orbiting stars other than our own Sun. The first exo-planet was discovered in 1992 by astronomers studying a section of the sky 2300 light years from the Sun. Astronomers use powerful telescopes to discover exo-planets, which are very hard to detect because they do not emit their own light and are very small compared to stars.

⟳ Keep up to date!

Many of the recent exo-planet discoveries have taken place using the Kepler space observatory.

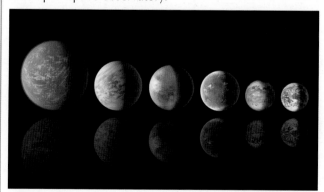

Figure 18.13 *An illustration comparing the size of some of the planets discovered by Kepler to Earth.*

Use NASA's Kepler website to find out how many exo-planets have now been discovered.

GO! Activity 18.4

☻ The challenges of living on Mars were explored in the 2015 Matt Damon movie *The Martian*. By referring to the requirements for a planet to be habitable, write a short essay on the challenges humans would face if a settlement on Mars were to be built.

🔍 Hint

Think about the conditions that exist on Earth.

Many of the exo-planets discovered exist within the habitable zones of the stars they orbit. In August 2016, observations made with a telescope in Chile detected a possible planet orbiting Proxima Centauri, the nearest star to the Sun. This planet, known as Proxima Centauri b, is believed to exist within the habitable zone. However, given the huge distance to Proxima Centauri, travelling to this planet using current space travel technology is not viable.

GO! Activity 18.5

1. Some exo-planets orbit stars in an area known as the habitable zone. State what is meant by the habitable zone.

2. There are billions of stars in the galaxy, and millions of galaxies in the universe. Explain what impact this may have on the likelihood of life existing elsewhere in the universe.

Learning checklist

After reading this chapter and completing the activities, I can:

N3 L3 N4 L4

- identify relationships between different properties of planets within the solar system by drawing and interpreting graphs. **Activity 18.1**   ○ ○ ○

- describe the following terms: planet, moon, star, Solar System, exo-planet, galaxy and universe. **Activity 18.2 Q1–Q3**   ○ ○ ○

- define a light year as the distance light travels in one year. **Activity 18.2 Q4**   ○ ○ ○

- convert distances between light years and metres. **Activity 18.2 Q5**   ○ ○ ○

- give examples of methods used to observe and explore space. **Activity 18.3 Q1**   ○ ○ ○

- give examples of the impact that space observation and exploration has had on our understanding of the universe and planet Earth. **Activity 18.3 Q3**   ○ ○ ○

- identify the conditions required for an exo-planet to sustain life. **Activity 18.4**   ○ ○ ○

- produce a reasoned argument for the likelihood of life existing elsewhere in the universe. **Activity 18.5 Q2**

Unit 3 practice assessment

N3 Forces

1. What is used to measure forces?   1

2. State the three things forces can do to objects.   1

3. A girl cycles along a path at a constant speed of 4 m/s. The total friction force acting on her and the bike is 100 N.

   Work out the forward force provided by the girl pedalling.   1

Cycling at a constant speed.

4. Give one example of where friction is useful and one example of where friction causes a problem.   2

N3 Solar System

5. The diagram below represents the different layers of the Earth.

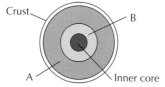

   (a) Name the layers represented by A and B.   2

   (b) Describe how the temperature of the Earth varies between the crust and the inner core.   1

6. Read the following article from a website which describes the formation of a volcano, and answer the questions.

   Cracks in the Earth's crust allow magma to rise up through them. This causes pressure to build up inside the Earth, which is released when tectonic plates move. The magma explodes onto the surface of the Earth. The lava flow eventually cools to form a new crust of igneous rock. This process repeats, building up layers of rock that form the volcano.

   (a) What causes the magma to explode onto the surface of the Earth?   1

   (b) What major rock type is formed when the lava cools to form a new crust?   1

7. The average temperature of the Earth has been rising for the past 100 years. State the name given to this change in the Earth's climate.   1

8. Day and night is caused because the Earth spins on its axis.

   (a) State how long it takes the Earth to spin on its axis.   1

   (b) Sundials were used in the past to tell the time. What part of the day is the shadow from a sundial at its shortest?   1

9. The phases of the Moon describe what the Moon looks like from the Earth as it orbits around the Earth.

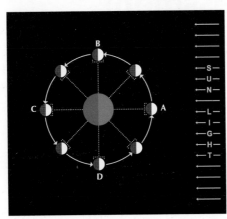

Name the phases of the moon at positions A, B, C and D. **4**

10. The Solar System is made up of planets which orbit the Sun.

   (a) State how many planets there are in the Solar System. **1**

   (b) Each planet takes a different time to orbit the Sun.

   (i) State which planet takes the longest to orbit the Sun. **1**

   (ii) State how long it takes the Earth to orbit the Sun. **1**

11. The number of hours of sunlight in a day changes throughout the year. The graph below shows how the length of day varies with the time of year.

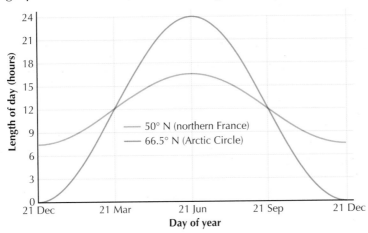

   (a) Estimate how much longer the length of a day is in northern France on 21st December compared with in the Arctic Circle. **1**

   (b) Estimate how much longer the length of a day is in the Arctic Circle on the 21st June compared with in northern France. **1**

   (c) What can be said about the day length in both places on 21st March and 21st September? **1**

National 4 Outcomes

N4 Speed and acceleration

1. At a school sports day you have been asked to calculate the average speed of runners during the 100 m, 200 m and 400 m races. Describe how you would measure the average speed of a runner. Your answer should include the measurements you would make and how you would make them, and any calculations you would do.   **3**

2. A pupil cycles 240 metres in 12 seconds. Calculate the average speed.   **3**

3. A car travelling 20 m/s drives for 50 m. Calculate the time taken for this journey.   **3**

4. What is the difference between average speed and instantaneous speed?   **2**

5. A rollercoaster is travelling at 4 m/s. Brakes are applied to bring the ride to a halt. The graph below shows how the speed of the ride changed.

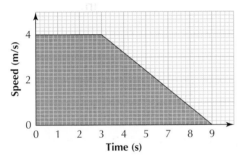

    (a) What is the total stopping distance of the ride?   **3**

    (b) Calculate the deceleration of the ride between 3 and 9 seconds.   **3**

    (c) A rollercoaster enthusiast is comparing roller coasters and finds the top speed.

Roller coaster	Ridesky	Watersplash	Fire
Top speed (m/s)	15		28

    Knowing that Watersplash is the slowest of the rides, suggest a value for its top speed.   **1**

N4 Forces

6. Calculate the force required to accelerate a car of mass 900 kg at 5 m/s².   **3**

7. Give a reason why a car will have a smaller acceleration in similar driving conditions when it tows a caravan.   **1**

8. Explain what happens to your weight as your travel away from the Earth's surface into outer space.   **1**

9. Use the information in the table to answer the following questions.

The effect of different values of gravity on the Moon and other planets in the Solar System.

	Earth	Moon	Mercury	Venus	Mars
Surface gravity (compared with Earth)	1.00	0.17	0.38	0.90	0.38
Your mass (compared with your mass on Earth)	1	1	1	1	1
How much you can lift (kg)	10	60	30	10	30
How high you can jump (cm)	20	120	53	22	53
How long it takes to fall back to the ground (s)	0.4	2.4	1.1	0.4	1.1

(a) State where the gravity is highest. **1**

(b) List the planets and the Moon in order of increasing gravitational attraction you would experience if you were standing on the surface. **1**

(c) Explain why the mass of an object is the same on all planets and on the Moon. **1**

10. An apple has a mass of 0.08 kg. What is its weight on Earth? ($g = 9.8$ N/kg) **3**

11. What forces are acting on a space shuttle during its re-entry into the Earth's atmosphere? **2**

N4 Satellites

12. Television signals are sent from America to the UK using geostationary satellites.

(a) State what is meant by a geostationary satellite. **1**

(b) The satellite dish receiver on the ground station is curved. Explain why the receiver is curved instead of flat. **2**

(c) A geostationary satellite is 35 800 km above the surface of the Earth. Calculate how long it takes the television signal to travel from the satellite to the ground station. **3**

13. Information about satellites orbiting at different altitudes is shown in the table below.

Satellite	Altitude (km)	Period of orbit (hours)	Speed of orbit (km/s)
A	180	1.5	7.8
B	2400		3.9
C	20 000	12	
D	36 000	24	3.1

    (a) State which satellite is a polar orbiting satellite.   **1**

    (b) Suggest a possible period of orbit for satellite B.   **1**

    (c) Suggest a possible speed of orbit for satellite C.   **1**

N4 Cosmology

14. State what is meant by a galaxy.   **1**

15. The distance to the Andromeda galaxy is 2.5 million light years. Explain what is meant by 2.5 million light years.   **1**

16. The following excerpt on exo-planets is found in a high school textbook:

An exoplanet is a planet which is not a part of our own Solar System. There are several methods that can be used to detect an exoplanet. One of these is the transit method where astronomers detect very small changes in the brightness of stars. When a planet passes in front of its star, it causes the light level from the star to drop. The changes in brightness can be observed over the course of years to prove that a planet is orbiting a star.

    (a) State what is meant by an exo-planet.   **1**

    (b) What can happen to the light level detected from a star when a planet passes in front of it?   **1**

    (c) State the name of the method for detecting exo-planets that is described in the passage.   **1**

17. The figure below shows how the period of orbit around the Sun is affected by the distance a planet is from the Sun.

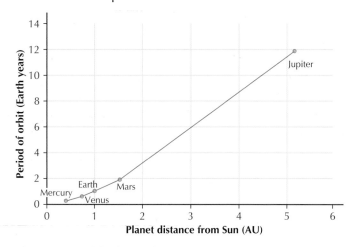

The asteroid belt is about 3 AU (astronomical units) from the Sun.

Use the graph to estimate the period of orbit of an asteroid around the Sun.   **1**

Investigation Skills

Planning and designing investigations • Practical skills • Data handling
• Presenting scientific findings

Appendix
Investigation
skills

Investigation skills

Planning and designing investigations

Learning intentions

In this section you will:

- learn that there are different parts to a scientific investigation
- identify the difference between independent, dependent and control variables
- learn that safety procedures are necessary when planning investigations.

Figure A.1 *An electricity experiment.*

📖 Word bank

- **Aim**

An aim is the question to be answered, or what is going to be investigated in an experiment.

- **Hypothesis**

A hypothesis is what you think will happen in an experiment.

- **Method**

The method is the set of instructions to be followed when doing an experiment.

- **Independent variable**

An independent variable is the variable that is changed or controlled in a scientific experiment.

- **Dependent variable**

A dependent variable is the variable being tested and measured in a scientific experiment.

- **Control variables**

Control variables are the variables that have to be kept the same each time the experiment is performed.

An experiment or investigation is used to find out what happens when something is changed. They are used by scientists to test their theories and to find answers.

Science experiments and investigations are used to answer questions about the way things behave. Investigations start with an **aim**, or a question you want answered.

Sometimes you might already have an idea as to what the answer might be. This idea is called a **hypothesis**. Scientists follow set **methods** to find out answers and test their hypothesis. These methods are like the recipe for baking a cake. Things have to be done in the right order to get the correct results. When you have to write a method, make sure you include all the instructions so that someone else could follow them and do the same experiment. It is usually helpful to include a diagram or photograph so people can see how you set up your apparatus. Make your diagrams clear and remember to label the different equipment.

Investigations usually involve measurements or observations made when changing one thing affects another thing. These are the **independent** and **dependent variables**:

- An **independent variable** is changed or controlled to test the effects on the **dependent variable**.

- The **dependent variable** is the variable being tested and measured in a scientific experiment.

There are also variables that have to stay the same each time the experiment is conducted. These are called **control variables**. In an investigation you can only change one thing at a time; this makes an experiment fair.

In most science experiments, the change in the dependent variable is measured using a range of suitable apparatus, depending on what you are trying to measure. Rulers, stopwatches, measuring cylinders and voltmeters can all be used to measure values in experiments. They all measure different quantities with different units. If you are planning your own investigation, think carefully about what you are trying to measure and the best piece of apparatus you could use to do so.

Example A.1

A simple physics experiment is planned to investigate the speed of a toy car as it travels down a slope. The method describes how to change the height of the slope by adding more blocks at one end.

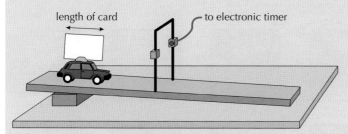

length of card   to electronic timer

What are the dependent, independent and control variables in this experiment?

The dependent variable is the speed of the car.

> The speed of the car is the variable being tested and measured.

The independent variable is the angle of the slope.

> The angle of the slope can be controlled or changed to test the effects on the control variable.

The control variables are: use the same car, use the same ramp, measure the speed at the same place.

> The car, the ramp and the place where the speed is measured all have to stay the same each time the experiment is carried out.

During an experiment you will have to record your observations or collect results. A table is a useful method to collect these in an organised way. Make sure you put the name of the measurement and its units in the heading row. Once data has been collected the aim can be considered. Has the question been answered? Is there a relationship (a link) between the variables? This is called a **conclusion**.

After an experiment has been conducted, an **evaluation** is written. The evaluation should consider how well the experiment worked and whether anything could be done differently if the test was to be repeated. Good evaluations can help experiments become more accurate or reliable.

📖 Word bank

• **Conclusion**

A conclusion is a short explanation stating what was found out in the experiment. It should answer the question set in the aim.

• **Evaluation**

An evaluation is a short explanation of the improvements that could be made to an investigation.

Safety precautions

Always follow instructions in experiments and investigations carefully. Read the instructions before starting the experiment. It may help to read the instructions a few times with a classmate to make sure everyone knows what to do. If you are not sure about something, always ask your teacher.

Make sure you follow all of the safety rules as you would normally be expected to do when carrying out practical work. You will often work in a small group but should record your own observations.

If you are asked to design your own experiment, make sure you think about the safety precautions you are going to take, and include them in your planning.

Figure A.2 *Science laboratory hazard signs.*

Practical skills

Learning intentions

In this section you will:

- explain the importance of taking accurate and reliable measurements
- learn that tables are used to clearly present data
- describe how to calculate an average.

You must always record your observations and measurements during an experiment. Your observations and measurements are often called **data**. You will need to measure these carefully and to record them in an organised way.

Accuracy and reliability

In an experiment you want to be as **accurate** as possible, to find the true value for your measurements. This means measuring things as carefully as possible. You can improve the accuracy of your results by choosing the correct apparatus for measurements, and by learning how to use the apparatus properly. Take your time when taking measurements. If they have small scales, get closer to the apparatus and read them at eye level. If the apparatus is faulty or if you make a mistake in the method, your results will be inaccurate.

Your experiments also need to be **reliable**. Data can be said to be reliable if measurements in the same experiment are repeated with no or only small differences each time the experiment is repeated.

Repeating an experiment and calculating average values is a way to improve your reliability. In this book it is suggested to repeat an experiment three times before finding the average value.

📖 Word bank

- **Accuracy**

A more accurate experiment will be closer to finding the true value of the measurement.

- **Reliability**

By repeating results and finding values that are approximately the same, the results can be said to be reliable.

Accuracy low – not in the centre
Reliability high – all in the same place

Accuracy high – all at the centre
Reliability high – all in the same place

Accuracy medium – nearer the centre
Reliability low – not in the same place

Accuracy low – not in the centre
Reliability low – all in different places

Figure A.3 *Accuracy and reliability.*

GO! Activity A.1

1. Copy and complete this list of some of the safety procedures you should follow in a laboratory, using the words provided (words may be used more than once). Your school may have more lab rules. Make sure you follow them.

   accidents   instructions   touch   ask   experiments

   Do not _ _ _ _ _ any equipment or supplies until clear _ _ _ _ _ _ _ _ _ _ _ _ have been given for you to do so and you understand the instructions.

   Report _ _ _ _ _ _ _ _ _ or spills immediately.

   Follow all _ _ _ _ _ _ _ _ _ _ _ _, they may be different for different _ _ _ _ _ _ _ _ _ _ _.

   If you don't know what to do, _ _ _.

   Do not run, jump, or jog in a laboratory as it may cause accidents.

2. What is a hypothesis?

3. Why do you need a diagram when setting up a scientific experiment?

4. Describe the difference between independent and dependent variables?

5. In an experiment you are asked to measure the length of a football pitch. Which of the following apparatus would you use and why?

   15 cm ruler, a metre stick, a tape measure or a trundle wheel

6. How can you make an experiment more accurate?

7. How can you make experimental data more reliable?

Tables and averaging data

A table is usually a good way to organise the information collected from an experiment.

- Organise the table with clear column headings.

- Always put the units of measurement in the heading.

- Usually, the independent variable goes in the left-hand column and the dependent variable goes in the right-hand column.

- Try to record your values to the same number of significant figures.

- You can add columns to the table for repeated measurements and averages.

Make the link

Significant figures are explained on page 377.

Example A.2

The table shows data collected from an experiment which measured the rebound height of a ball from different surfaces.

Surface	Rebound height (cm)			Average rebound height (cm)
	test 1	test 2	test 3	
Carpet	23	24	22	
Wood	105	113	100	
Concrete	125	129	121	
Rubber mat	45	54	48	

Calculate the average rebound height for balls on the carpet surface.

> The type of surface is the independent variable and goes in the left-hand column.

> The rebound height is the dependent variable and goes in the right column-hand column.

Surface	Rebound height (cm)			Average rebound height (cm)
	test 1	test 2	test 3	
Carpet	23	24	22	23
Wood	105	113	100	
Concrete	125	129	121	
Rubber mat	45	54	48	

> The table has three columns for three different sets of experiments. The measurements are repeated and the average calculated. This improves the **reliability** of the experiment.

test 1 = 23 cm

test 2 = 24 cm

test 3 = 22 cm ——— (Write what you know from the question.)

average = ? ——— (Write what you are being asked to find.)

$$\text{average} = \frac{(\text{test 1} + \text{test 2} + \text{test 3})}{\text{number of tests}}$$ ——— (Write the equation you need to use.)

$$\text{average} = \frac{(23 + 24 + 22)}{3}$$ ——— (Substitute what you know.)

average = 23 cm ——— (Calculate the value. Don't forget to write the units in your answer.)

GO! Activity A.2

Copy the table for Example A.2 and find the missing values for the average rebound height for the other surfaces.

Data handling

Learning intentions

In this section you will:

- learn how to set out numerical calculations
- learn that values should be given to an appropriate number of significant figures
- explain the differences between pie charts, bar charts and line graphs
- learn how to draw pie charts, bar charts and line graphs
- learn that there can be a pattern or trend in data collected
- describe how to draw a line of best fit on a line graph.

Figure A.4   *Ready to handle data?*

Data handling is an essential part of the scientific process. Data handling involves:

- recording data
- calculating with data
- analysing data and calculations
- communicating results and analysis.

In science, diagrams, charts and graphs are often used to communicate information. Using standard methods of communication makes it easier for other scientists to understand what has been done.

Calculations

Many experiments in science, and in physics in particular, will involve numerical calculations. The calculations will often use the relationships described in each chapter.

N3   L3   N4   L4

When you are carrying out the calculations, it is important to set out your work neatly. This makes it easier to work out what to do, and reduces the chances of making careless mistakes. In this book, we use this sequence of steps for numerical calculations using relationships:

- write out what you know and what you are being asked to find
- write out the equation (relationship)
- substitute in what you know
- calculate the value
- don't forget to write the units with your answer.

Equations in physics usually use symbols instead of words. For example, the equation

$$\text{speed} = \frac{\text{distance}}{\text{time}}$$

is usually written as

$$v = \frac{d}{t}$$

where:

v represents speed

d represents distance

t represents time.

Using symbols or letters happens in other subjects as well. For example, in maths, Pythagoras' theorem can be written as $a^2 + b^2 = c^2$, where the letters represent the lengths of the sides on a right-angled triangle.

Always check that you have answered exactly what the question has asked. It is also important to check your final answers to see if they make sense. Sometimes you can spot an error in your calculation just by noticing that your answer does not make sense.

It can be useful to show your final answer using a prefix. A prefix goes before the unit of measurement help write very large or very small numbers.

Table A.1   *Table of prefixes.*

mega	M	1 000 000	million
kilo	k	1000	thousand
milli	m	1/1000	thousandth
micro	μ	1/1 000 000	millionth

Example A.3

A runner travels 100 metres in 10 seconds. Calculate his average speed.

distance = 100 m

time taken = 10 s ● —————— Write out what you know from the question and what you are being asked to find.

average speed = ?

average speed = $\dfrac{\text{distance}}{\text{time}}$ ● —————— Write out the equation.

average speed = $\dfrac{100}{10}$ ● —————— Substitute in what you know.

average speed = 10 m/s ● —————— Don't forget to write the units in your answer.

Example A.4

A ball has an average speed of 0.13 m/s. Calculate how far it will travel in 6 seconds.

$v = 0.13$ m/s

$t = 60$ s ● —————— Write out what you know from the question and what you are being asked to find.

$d = ?$

$v = \dfrac{d}{t}$ ● —————— Write out the equation.

$0.13 = \dfrac{d}{6}$ ● —————— Substitute in what you know.

distance = 0.13 × 6

distance = 0.78 m ● —————— Don't forget to write the units in your answer.

GO! Activity A.3

1. Use the relationship average speed = $\dfrac{\text{distance}}{\text{time}}$ to complete the missing values **(a)–(e)** in the table.

Distance (m)	Time (s)	Speed (m/s)
20	5	**(a)**
1000	10	**(b)**
(c)	9	81
5000	**(d)**	50
75	**(e)**	25

Significant figures

It is very important to give the answer to an appropriate number of **significant figures**. The significant figures relate to the precision of measurements and calculations.

In any given number, the figures 0 to 9 are deemed to be **significant** if:

- they are not a zero. (Any number from 1 to 9.)
- they are a zero that is sandwiched between other non-zero figures. (For example, the zero in the number 302 is significant because it tells you that there are no tens in the number, and it determines the place value of the 3.)
- they are a zero that comes after the decimal place. (For example, the number 5.0 has two significant figures; the zero is significant because it tells you there are no tenths. Otherwise it would not be needed.)

There is a simple rule for rounding final answers to calculations:

> **Round the final answer to the same number of significant figures as in the question.**

GO! Activity A.4

State how many significant figures there are in the following numbers:

1. 60300
2. 3.1402×10^8
3. 3.00×10^8
4. 314.15
5. 0.002468

Charts and graphs

Charts and graphs are often used to display data. They make it easier to find trends or patterns in your results. The type of data you collect in the investigation or experiment might be best drawn as a chart or as a graph.

Pie charts

A pie chart is a good way to display data that can be shown as percentages or fractions of a whole. The size of each segment represents the relative size of each piece of data.

Use a protractor to measure the angles for each segment. Remember to start at the top position (at 12 o'clock) and draw the biggest piece of the pie chart first. Add each smaller piece going clockwise round the circle. Each segment must be labelled with the name and the value of what it represents.

Example A.5

The table shows different sources of background nuclear radiation. Show the data in a pie chart.

Source of background radiation	Percentage
Radon gas	50.0
Buildings and the ground	14.0
Medical	14.0
Food and drink	11.5
Cosmic rays	10.0
Nuclear power and weapons tests	0.3
Other	0.2

Source of background radiation	Percentage	Angle
Radon gas	50.0	$50\% \times 360° = 180°$
Buildings and the ground	14.0	$14\% \times 360° = 50.4°$
Medical	14.0	$14\% \times 360° = 50.4°$
Food and drink	11.5	$11.5\% \times 360° = 41.4°$
Cosmic rays	10.0	$10\% \times 360° = 36°$
Nuclear power and weapons tests	0.3	$0.3\% \times 360° = 1.08°$
Other	0.2	$0.2\% \times 360° = 0.72°$

To show the data in a pie chart, the percentages must be expressed in degrees. A whole circle has 360°, so the angles for each sector are calculated as fractions of 360°.

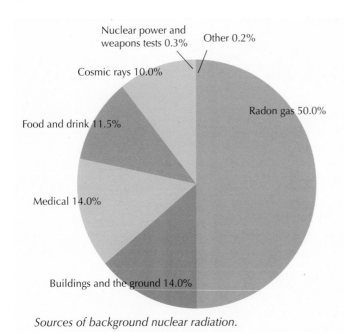

Sources of background nuclear radiation.

GO! Activity A.5

1. Use the information in this table to draw a pie chart.
2. What do all the percentages add up to?

Percentage of houses using different fuels.

Type of fuel	% of houses using the fuel
Wood	30
Gas	50
Electricity	10
Coal	5
Peat	5

Bar charts

Bar charts are used to represent numbers that are not percentages or fractions. They are used when experiments have different categories for the independent variable **(discrete data)**. When you draw a bar chart:

- choose a scale that means you can fit every number on your graph paper
- make sure your scale is regular, like the markings on a ruler, and starts at zero
- try to use as much space as you can – don't make the graph too small
- draw and label each axis carefully, and include units in the labels where necessary
- make the bars the same width and leave equal spaces between the values
- keep the top of each bar straight – use a ruler
- label each bar with the name of what it represents.

📖 Word bank

- **Discrete data**

Discrete data can only take certain values and is based on counted values (such as students in a school). The values cannot be divided meaningfully (you can't have half a student).

Example A.6

The table shows the densities of different materials. Show this data in a bar chart.

Material	Density (g/cm³)
Ice	0.92
Liquid water	1.00
Rubber	1.50
Aluminium	2.70
Wood	0.70

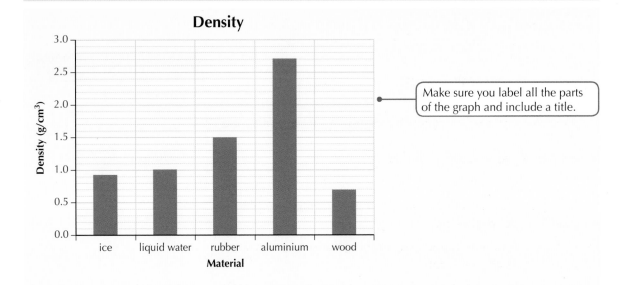

Make sure you label all the parts of the graph and include a title.

GO! Activity A.6

1. The table shows the frictional forces on different surfaces. Use the information in this table to draw a bar chart.

2. Using the chart, which surfaces have the same height bar?

3. How many times greater is the value of the frictional force on carpet than on wood? What does that mean for the height of the bars?

Surface	Frictional force (N)
Wood	2
Carpet	4
Sandpaper	5
Polished floor	2

Line graphs

Line graphs are drawn when there is **continuous data** for both the dependent and independent variables. All the results have numerical values.

Line graphs are very useful in showing trends or patterns in results. They can also be used to make predictions or forecasts.

Points must be plotted carefully. A **line of best fit** can be drawn to show patterns. This is sometimes called a **scatter graph**.

Normal practice is to show the independent variable on the x-axis (horizontal) and the dependent variable on the y-axis (vertical).

When drawing a line graph, remember to:

* choose scales that mean you can fit every number on your graph paper
* make sure your scale is regular, like the markings on a ruler, and starts at zero
* label each axis with the variable being measured and its units
* plot each point using a × or a ⊙.

Sometimes you will be asked to join the data points together, to show a trend, or to draw a line of best fit through the points to show the pattern they make.

Do not extend the line to the origin unless there is data with a value of 0 for both the variables.

In many physics topics, you will be asked to find the relationship between two variables. Use a line of best fit. The horizontal and vertical axes should start at zero. The line of best fit may not join all the points, but shows the line which fits best to all the plotted points. A line of best fit might not be a straight line. Depending on the relationship being shown, the line of best fit might be a curve.

Lines of best fit can be used to predict values for which you don't have data, by extending the line past the data points collected. Care needs to be taken with extending the line in case the pattern does not continue.

Word bank

* **Continuous data**

Continuous data can have any numerical value (usually within a range).

* **Line of best fit**

A line of best fit is a line which shows the relationship or pattern on a graph. It can be a straight line or a curve.

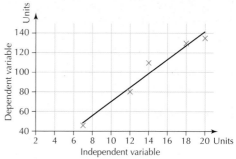

Figure A.7   *A line graph showing a trend.*

A line graph will let you spot **anomalous** data. It is always a good idea to think about any anomalous data, because it might tell you something about the experiment or calculations you have used. Was there a mistake in the experiment? Did all the apparatus work properly?

The slope of the straight line is the **gradient**. Gradients are a measure of **rates of change**. The gradient tells you how fast something is changing. In a speed–time graph the gradient represents the acceleration of the object.

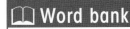

Word bank

• **Anomalous**

Data that looks out of place is called anomalous data. This could be due to a mistake in the graph or an error in the experiment.

Example A.7

The table shows the speed of a car at different points in the journey. Use this data to draw a line graph.

Time (s)	0	2	4	6	10
Speed (m/s)	0	4	8	12	20

Plot each point using the data from the table, and join with a straight line.

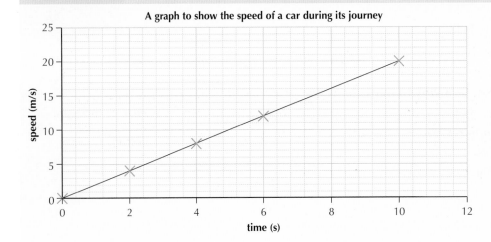

A graph to show the speed of a car during its journey

GO! Activity A.7

1. The table shows the temperature of a beaker of water as the water cools from boiling point. Use the information in this table to draw a line graph. Remember to label the axis with the correct units.

Time (min)	Temperature (°C)
0	100
1	80
2	65
3	60
4	43
5	40
6	29
7	20
8	12
9	?

2. Draw a straight line of best fit to show the relationship between temperature and time.

3. What temperature do you think the water would be at 9 minutes?

4. Can you explain why all the points don't lie on your line of best fit?

 Hint

This experiment has only been done once, so these are not averages. Think about errors in collecting the data.

Presenting scientific findings

Fig A.9 *Displaying scientific reports as posters.*

Fig A.10 *A presentation of an investigation.*

Fig A.11 *Writing a report in a group.*

There are many different ways to share your scientific findings. You will need to think about:

- who is going to see it
- what is the best way to display the information.

Sometimes you might present the same information in different ways for different people.

You might write a detailed report explaining how you did an experiment, with diagrams and charts, for your classmates, and you might also display your data using posters with colourful pictures and large graphs for visitors to the school.

It is very important to summarise the data and the information so your report is not too long. You should also **reference** any supporting material you have used. When you present your findings be sure to acknowledge all your sources, whether they are library books, websites, or conversations with teachers and students. References are included so that you can acknowledge the contribution of others' research in your work.

Acknowledging sources

Forms of references could be:

- A book reference: Devine, S; McLean, D; Smith, S. (2014) National 5 Physics Student Book, Leckie and Leckie, p104
  (Write as much detail as you know about the book so that someone else could find it and the page you looked at.)

- A person you have spoken or listened to: A presentation by J Spence and A Lee on Jan 17 2017. (Put the names of the people and the date.)

- An internet page: http://www.educationscotland.gov.uk/learnerresources/index.asp, 1/11/16. (Put the full URL and the date you accessed the page.)

GO! Activity A.8

1. List some ways to share the results of your experiment with others.
2. List three places you could find information about the different planets in our Solar System.
3. Search online or use another research method, find out the diameter of the Moon. Write your answer and include your reference.

Learning checklist

After reading this section and completing the activities, I can:

N3 L3 N4 L4

	• state that there are different parts to a scientific investigation. **Activity A.1 Q2, Q3**	○	○	○
	• state the difference between independent, dependent and control variables. **Activity A.1 Q4**	○	○	○
	• state that safety procedures are necessary when planning investigations. **Activity A.1 Q1**	○	○	○
	• explain the importance of taking accurate and reliable measurements. **Activity A.1 Q5–7**	○	○	○
	• state that tables are used to clearly present data. **Activity A.2**	○	○	○
	• describe how to calculate an average. **Activity A.2**	○	○	○

N3 L3 N4 L4

- give an example of how to set out numerical calculations. **Activity A.3**

- state that values should be given to an appropriate number of significant figures. **Activity A.4**

- give examples of when to draw pie charts, bar charts and line graphs. **Activity A.5, Activity A.6, Activity A.7**

- state that there can be a pattern or trend in data collected. **Activity A.7**

- describe how to draw a line of best fit on a line graph. **Activity A.7**

- state that two examples of how scientific findings can be shared are presentations and posters. **Activity A.8 Q1**

- name and identify some of the different places scientific information can be collected from. **Activity A.8 Q2**

- state that all sources should be acknowledged. **Activity A.8 Q3**

© 2017 Leckie & Leckie Ltd

001/21062017

10 9 8 7 6 5 4 3 2 1

ISBN 9780008204495

Published by
Leckie & Leckie Ltd
An imprint of HarperCollinsPublishers
Westerhill Road, Bishopbriggs, Glasgow, G64 2QT
T: 0844 576 8126   F: 0844 576 8131
leckieandleckie@harpercollins.co.uk
www.leckieandleckie.co.uk

Commissioning editor: Clare Souza
Managing editor: Craig Balfour

Special thanks to
Janet Marriot (copy edit)
Jess White (proofread)
Project One Publishing Solutions (project management and editorial)
Jouve India (layout and illustration)
Ink Tank and Ian Wrigley (cover design)

The publishers would like to thank David and Maheswari of Jouve India for their boundless spirit, unrivalled dedication and relentless effort throughout this publication.

Printed in Italy by Grafica Veneta SpA

A CIP Catalogue record for this book is available from the British Library.

Acknowledgements

Cover photograph of British ESA Astronaut Tim Peake. Copyright: ESA/NASA.

Figure 1.2 © stocker1970 / Shutterstock.com; Figure 1.4 © Grandpa / Shutterstock.com; Figure 1.5 © CappaPhoto / Shutterstock.com; Figure 1.6 © Diyana Dimitrova / Shutterstock.com; Figure 1.11a © oleandra / Shutterstock.com; Figure 1.11b © LI CHAOSHU / Shutterstock.com; Figure 1.11c © Yury Stroykin / Shutterstock.com; Figure 1.14 © Stanley Howe and is licensed for reuse under the Creative Commons Attribution-ShareAlike 2.0 license; Figure 1.17 © MISS KANITHAR AIUMLA-OR / Shutterstock.com; Figure 1.21 © Olga Marc / Shutterstock.com; Figure 1.22 © REUTERS / Alamy Stock Photo; Acticity 1.11 © Ahctiqus Lee / Shutterstock.com; Figure 1.25 © Mike Garrett and licensed under the Creative Commons Attribution 3.0 Unported license; Figure 2.1 © SimonHS / Shutterstock.com; Figure 2.4 © Bplanet / Shutterstock.com; Figure 2.6 © SamJonah / Shutterstock.com; Figure 3.1 © revers / Shutterstock.com; Figure 3.2 © Aleksandr Pobedimskiy / Shutterstock.com; Figure 3.3 © Adam Fraise / Shutterstock.com; Figure 3.4 © Zoltan Gabor / Shutterstock.com; Figure 3.5 © Dorling Kindersley/UIG/SCIENCE PHOTO LIBRARY; Figure 3.6 © snapgalleria / Shutterstock.com; Figure 3.7 © CORDELIA MOLLOY/SCIENCE PHOTO LIBRARY / Shutterstock.com; Figure 3.8 © MilanB / Shutterstock.com; Activity 3.1a © MilanB / Shutterstock.com; Figure 3.11 © Drummond Fyall, Created Eye Photography Figure 3.12 © Africa Studio / Shutterstock.com; Figure 3.13 © Dr_Flash / Shutterstock.com; Figure 3.14 © Amnarj Tanongrattana / Shutterstock.com; Acticity 3.4 © Fouad A. Saad / Shutterstock.com; Figure 3.18 © SvedOliver / Shutterstock.com; Figure 3.20 © Designua / Shutterstock.com; Figure 3.23 © cyo bo / Shutterstock.com; Figure 3.24 © EPSTOCK / Shutterstock.com; Figure 3.25 © CGinspiration / Shutterstock.com; Figure 3.26 © alpinethread and licensed under the Creative Commons Attribution-Share Alike 2.0 Generic license; Figure 3.29 © photoiconix / Shutterstock.com; Activity 3.9 © © photoiconix / Shutterstock.com; Figure 4.1 © Jakinnboaz / Shutterstock.com; Figure 4.3 © ODI / Alamy Stock Photo; Figure 4.4 © sciencephotos / Alamy Stock Photo; Figure 4.5 © Simon Bratt / Shutterstock.com; Figure 4.6 © Scanrail1 / Shutterstock.com; Figure 4.9 © SCIENCE PHOTO LIBRARY; Figure 4.10 © TREVOR CLIFFORD PHOTOGRAPHY/ SCIENCE PHOTO LIBRARY; Figure 4.14 © oliveromg / Shutterstock.com; Figure 4.17 © IB Photography / Shutterstock.com; Figure 4.19 © HardheadMonster / Shutterstock.com; Figure 4.20 © TREVOR CLIFFORD PHOTOGRAPHY/SCIENCE PHOTO LIBRARY; Figure 4.23 © Volodymyr Krasyuk / Shutterstock.com; Figure 4.26 © nobelio / Shutterstock.com; Figure 4.27 © Chones / Shutterstock.com; Figure 4.28 © Apinan / Shutterstock.com; Figure 5.1 © George Burba / Shutterstock.com; Figure 5.4 © Designua / Shutterstock.com; Figure 5.5 © BlueRingMedia / Shutterstock.com; Figure 5.6 © saerdnarelleumrebo / Shutterstock.com; Figure 5.7 © Roman Vukolov Shutterstock.com; Figure 5.8 © Taina Sohlman / Shutterstock.com; Figure 5.9 © CHARLES D. WINTERS/SCIENCE PHOTO LIBRARY / Shutterstock.com; Figure 5.10 © SCIENCE PHOTO LIBRARY / Shutterstock.com; Activity 5.2 © Shutterstock.com; Figure 5.11 © MR.YURANAN LAKHAPOL / Shutterstock.com; Figure 6.4 © PhotographyByMK / Shutterstock.com; Figure 6.6 © dmytro herasymeniuk / Shutterstock.com; Figure 6.8 © Kichigin / Shutterstock.com; Activity 6.1 © Sascha Burkard / Shutterstock.com; Figure 6.10 © ThamKC / Shutterstock.com; Figure 6.11 © pefostudio5 / Shutterstockc.com; Activity 6.2 © Aleksandr Veremeev / Shutterstock.com; Figure 6.12a Quayside / Shutterstock.com; Figure 6.12b © JOHNNY GREIG/SCIENCE PHOTO LIBRARY; Figure 6.12c © Caron Badkin / Shutterstock.com; Figure 6.13 © Craig Russell / Shutterstock.com; Figure 6.14a © Mrs_ya / Shutterstock.com; Figure 6.14c © Paul Reid / Shutterstock.com; Figure 6.14d © Paul Reid / Shutterstock.com; Figure 6.15 © Gary Perkin / Shutterstock.com; Figure 6.17a © Swasdee / Shutterstock.com; Figure 6.17b © Sinchukov Aleksander / Shutterstock.com; Figure 6.17c © dmytro herasymeniuk / Shutterstock.com; Figure 6.17c © Claudio Divizia / Shutterstock.com; Figure 7.1 © Syda Productions / Shutterstock.com; Figure 7.3 © Shutterstock.com; Figure 7.5 © Rasulov / Shutterstock.com; Figure 7.6 © studiovin / Shutterstock.com; Figure 7.7 © MARTYN F. CHILLMAID/SCIENCE PHOTO LIBRARY / Shutterstock.com; Figure 7.8 © Timothy Hodgkinson / Shutterstock.com; Figure 7.9 © Carlos Munoz / Shutterstock.com; Figure 7.10 © ryabuha kateryna / Shutterstock.com; Figure 7.11 © ANDREW LAMBERT PHOTOGRAPHY/SCIENCE PHOTO LIBRARY; Figure 7.12 © Oleksandr Kostiuchenko / Shutterstock.com; Fig 7.13 © GIPHOTOSTOCK/SCIENCE PHOTO LIBRARY; Fig 7.15 © tantawat / Shutterstock.com; Figure 7.16 © Fouad A. Saad / Shutterstock.com; Figure 7.17 © pimpisan02 / Shutterstock.com; Figure 7.18 © Krasowit / Shutterstock.com; Figure 7.20 © Hurst Photo / Shutterstock.com; Figure 7.21 © Chones / Shutterstock.com; Figure 7.22 © moreimages / Shutterstock.com; Figure 7.23 © Warut Chinsai / Shutterstock.com; Figure 7.24 © Pongsak A / Shutterstock.com; Figure 7.26 © Sergiy Kuzmin / Shutterstock.com; Figure 7.28 © ZEF / Shutterstock.com; Activity 7.9 © ANDREW LAMBERT PHOTOGRAPHY/SCIENCE PHOTO LIBRARY; Figure 8.1 © Natalka Dmitrova; Figure 8.2 © Steve Allen / Shutterstock.com; Figure 8.4 © holbox / Shutterstock.com; Activity 8.3 © MARTYN F. CHILLMAID/ SCIENCE PHOTO LIBRARY; Figure 8.6 © sciencephotos / Alamy Stock Photo; Figure 8.7 © Fouad A. Saad / Shutterstock.com; Figure 8.8 © Designua / Shutterstock.com; Figure 8.10 © Kaesler Media / Shutterstock.com; Figure 8.11 © mediacolor's / Alamy Stock Photo; Figure 8.12 © NASA; Figure 8.13 © Slavo Valigursky / Shutterstock.com; Figure 8.14 © Monkey Business Images / Shutterstock.com; Figure 8.15 © Ivan Smuk / Shutterstock.com; Figure 8.16 licensed under the Creative Commons Attribution-Share Alike 3.0 license; Figure 8.18 © RMIKKA / Shutterstock.com; Figure 8.19 © Denis Dryashkin / Shutterstock.com; Figure 8.20 © WaitForLight / Shutterstock.com; Figure 8.25 © mehmet dinler / Shutterstock.com; Figure 8.26 © Sherry Yates Young / Shutterstock.com; Figure 8.27 © DJ Mattaar / Shutterstock.com; Figure 9.1 © Nicholas Rjabow / Shutterstock.com; Figure 9.3 © Georgre Gray / Shutterstock.com; Figure 9.8 © Vasin Lee / Shutterstock.com; Figure 10.2 ©